ICE THERAPY
NATURE'S PERFECT CURE®

BY Harry A. Thompson

Dedication

To the woman in the hot tub who insisted that I get out of the hot water because it was making my "hot, angry muscles hotter and angrier." She said that if I wanted to get better, I should buy hot/cold reusable gel packs and ice my sore back instead. She didn't tell me her name and I never saw her again.

If you're out there, this is for you.

Also, my book is dedicated to the pioneering people who have undertaken this treatment on their own to reduce their pain, and to the many more that will in the future.

And it is dedicated to the chiropractors, medical doctors, and physical therapists that use ice therapy to alleviate pain for their patients.

Special Thanks

To Amber Westbrook for her support from the very beginning when I was sure no one would believe me; to Dr. Kirstin Ellsworth, Ph.D., for her personal encouragement and literary know-how; and to Jan Fowler, for breathing life into a manuscript that was badly in need of resuscitation.

"Two years ago, following lung surgery, I developed a severe sciatic condition and was unable to get out of bed or walk. Oral medication was not an option due to a liver problem and, because of the recent surgery, I could not go to the recommended chiropractor. During this discouraging time, Harry shared his own icing theory and successful results with me. This easy, convenient and cost-free method was worth a try. After 'icing' for several weeks, I am now mobile and pain free. The 'best kept secret' is the DURATION of icing...at least 3 - 4 hours during one application. Try it...you won't believe the results!"

— Connie Mason, Apple Valley, CA

Foreword

Before I began ice therapy, I had tried every treatment and medication available for pain, and absolutely nothing had worked. As I began icing, I noticed that as long as I kept the ice in place, the inflammation and swelling of the connective tissue (muscles, tendons, ligaments, and cartilage, their attachments to bone and each other, and the nerves that energize them) continued to diminish; with gentle encouragement, the tissue and joints realigned themselves.

As time passed, I discovered that it was necessary to keep the cold in place for up to four hours in order for chronic injuries to respond. In the process, I switched from gel packs to an ice bag when I found that the ice bag could keep areas colder longer, and I could ice more often, because I didn't have to wait hours for it to refreeze.

Since stumbling onto this miracle of nature, I have iced away my pain, and my body is functioning normally. It's no secret that ice has been used in medicine for years to reduce inflammation and swelling. But it's the employment of the time factor described in this book that makes the difference: it unleashes a natural healing power that, until now, has been standing in the wings, waiting to be discovered.

If all the doctors and therapists who claim they know how to fix your pain actually did, if all the methods in all the books promising to rid you of pain actually could, and if all the medications, meditations, shocks, vibrators, salves, poultices, and potions hocked as miracle cures actually did what they claim, chronic pain would have been a thing of the past long ago. Take it from someone who's "been there, done that"—all of that. The definition of insanity is doing the same thing over and over, expecting a different result. Now you can stop doing what doesn't work and start doing what does.

In the following pages, I guide you through the icing process step by step with words and pictures, and tell you what to expect so you can locate your pain and ice it away. I have tried to use common terms rather than medical terms, and give definitions for the medical terms I must use, to make this book easier to understand.

By treating yourself in the privacy of your own home at your own pace, you'll avoid waiting rooms, co-pays, and insurance companies that deny you coverage. If you ice for more than forty-five minutes at a time and perform an easy exercise afterwards, you'll begin to feel better.

Four years into my ice therapy, I retired to Costa Rica and had to ship my belongings there; I realized that heavy, leaky gel packs were not an option. I was forced to switch to an ice bag and as a result I discovered the time factor. Ice bags stayed cold longer, and the "longer" was the key to recovery. I began to see results in leaps and bounds due to this time factor. It was then that the idea of sharing this information with others took root and I began treating my therapy as a job. I planned my days around icing and began writing down the results.

I didn't have every injury there is to have (thank goodness), but the injuries I did have ran from the top of my head to the bottom of my feet and almost everywhere in-between — so I had to work out how to alleviate pain in every area of my body.

Ice therapy (not to be confused with "cold therapy" machines) was a last resort for me and, to my utter amazement, it cured my aches and pains and every other ailment I was living with. Most astonishing of all is that ice therapy is, simultaneously, completely benign and incredibly powerful, given enough time. During the seven years that I treated myself with ice therapy, it only produced positive results.

Section I: The Conquest of Pain

"For several months I've been having consistent pain in and all around my left knee. When I would sit down for a while and then try to get up, my knee started to lock up and would not allow my leg to straighten out without shooting pain coming from the center of the knee and going all around the knee. It would keep me from putting any pressure on the leg to stand up.

I was telling Harry about my problem when he suggested that I take an ice bag filled with ice and put it on the area that was in pain. He insisted that I keep the ice bag on the painful area for at least three hours and to keep the bag drained and filled periodically in order to get the full effect of the cold treatment.

I definitely followed his directions because when you are in severe pain you will try just about anything just to get some relief. I followed his directions to a "T" and around the third hour I could barely stand the cold but to my shock and surprise, when I removed the ice pack from my knee and tried to stand up on my left leg, the pain had subsided enough that I could put pressure on it without the shooting pain and I'm now walking pain free. I repeated the process until there was no pain at all."

—Amber Westbrook, Lynnwood, WA

Chapter 1. See for Yourself

My story might be much like yours, only mine may be longer and more complicated because of the sheer number of accidents and work-related injuries I have accumulated. Maybe you and I are both lucky that I've been through all this, because my pain drove me to find a simple and non-invasive way to get rid of it.

Here's why I was in so much pain and why, later in life, I began experiencing a host of other physical ailments that I now know were directly or indirectly caused by these injuries:

- When I was four years old, I opened the back door of my mother's car and fell out. Fortunately, she was driving on city streets and wasn't traveling fast. A policeman saw it happen; he put me in his car and took me to the hospital. I had fainting spells for several years afterwards.

- When I was twelve, I walked up behind my horse and he kicked me in the stomach. I flew through the air and landed on my back. My stomach was sore for a week but I didn't break any bones and there were no internal injuries.

- When I was fourteen, I took a job with a rock mason. Every Saturday, we'd drive to a rock quarry a few miles away. I would climb a rock pile, my boss would point to the rocks he wanted, and I'd roll them down the hill and load them in the back of his pickup truck. For half a day, we'd gather enough rocks to keep him busy during the week while I was in school. I did that every Saturday for over a year. That was the beginning of my back problems.

- When I was nineteen, I crashed a motorcycle and broke my right ankle and fibula. The doctors operated and put a screw in the ankle and a pin in the leg. My helmet was cracked from the top down the right side and I was unconscious when I was taken to the hospital.

- In my early twenties, I was driving on the freeway and rear-ended the car in front of me. I was wearing my seatbelt and shoulder harness, but my knees were jammed under the dashboard. I was taken to the hospital and released with "no serious injuries." For weeks, both my knees were swollen and felt spongy to the touch, and my neck was stiff.

- From age twenty to thirty-five, I worked as an assistant cameraman and carried heavy cameras and equipment for workdays that lasted twelve hours or more.

- On a trip to New York City, I crossed 7th Avenue heading uptown and was hit by a car. I did a flip in the air and landed in the street on my back. I managed to break the fall with my left heel and both hands. I had cuts and bruises but no broken bones. I was treated at a hospital and released.

- In my mid-thirties, I was involved in an auto accident that put me in the hospital for two weeks. My right humerus (upper arm bone) was broken, my nose was broken, and I had cuts and bruises. Later I was told that the engine of the car was sitting in the driver's seat.

- I was an X-ray technologist for fourteen years, and piled injuries associated with health care on top of my old injuries. By the time I was fifty-six years old, I was in so much pain that I went back to school for a year to learn echocardiography, a job I could do sitting down and one that didn't require heavy lifting.

If you have experienced similar injuries, if you are suffering with chronic pain and standard treatments have been unsuccessful, ice therapy is the answer you have been waiting for.

I'm not a medical doctor, but I do have medical training. I've seen quite a few people trapped in unhandled pain. And I had my own pain as a powerful argument that pain cripples your life.

I'm an adventurous person — just look at all those injuries — and I have tried many things in life, including the adventure of facing and erasing my own pain.

You won't believe this book just because I say so, and you shouldn't. You need to see for yourself by trying it. Read just enough to try my discovery out and if it works for you, read more.

The next chapter tells you how to start.

Chapter 2. How to Use This Book

I wrote this book for you to use. I want you to be able to take the fastest route out of pain. The information is all here and I hope you'll read it all but you can approach it in any way you like. The purpose of this chapter is to give you a suggested approach, some simple steps to make it easiest to try ice therapy. Here's what I suggest:

a. Read the short chapter after this one; it will give you a foundation, some basic principles involved in ice therapy.
b. Pick a section of the body from the chapters that follow in the Where's the Pain section. Choose one that covers an area where you have pain.
c. Read the chapter on that section, so you can share my experience with handling pain in that area.
d. Go to the Method section and read those 3 short chapters so you know how to do this simple therapy.
e. Then, don't waste any time! Start to work on that area where you have the most pain and follow what I've done in that area as closely as you can. (After all, I went through all this and learned from my mistakes — you might as well benefit from my experience.) You can continue to read this book while you're icing if you want to.
f. If you can, keep a notebook or something to record what you do and the changes you experience as you use ice therapy.
g. If your pain gets better, and I am convinced it will, you can also:
   1. read more of this book.
   2. try other areas and alleviate the pain there, too.
   3. get another copy of this book and give it to a friend who has pain.

Chapter 3. What You Need to Know to Start

Hot and cold

The primary principle on which this therapy is based is the idea that for purposes of relieving pain, HEAT=BAD, and COLD=GOOD. I learned this from a lucky encounter with a woman in a fitness-center hot-tub, who insisted that I get out of the hot water because it was making my "hot, angry muscles hotter and angrier." She said that if I wanted to get better, I should buy hot/cold reusable gel packs and ice my sore back instead.

I was running out of ideas for handling my chronic pain when she offered this free advice, so I tried it. I would have tried almost anything. But she was right. In fact I found out something further by experimenting: that the longer an area stayed cold, the better. So an ice bag was even better than gel packs because it stayed cold longer and could be drained and refilled, allowing me to ice 24/7, if I had the ice and the will.

Palpating

Palpating means touching with your fingers or hands to see what your sense of touch tells you. You will be palpating different areas of your body to find locations where there is pain. Press hard, and dig deep, to find the source of your chronic pain.

The ice bag as a diagnostic tool

The ice bag, as well as helping to heal pain, is a diagnostic tool — that is, it helps you to find where your body isn't doing so well. Icing an area often exposes related pain in other areas, which I call referral pain. If you note these areas down you can remember to ice them later.

The "forty-five minute rule"

In icing, I discovered early on that about every forty-five minutes, a pain-releasing event would take place, either at the site of the pain, or in another area of my body, usually on the opposite side to the one I was icing. These events could include a burning sensation, a heightened sense of pain, muscles relaxing, or "electric" shock. These events were so regular that I began calling this phenomenon the "forty-five minute rule." It's amazing how regular this phenomenon is. I used the anticipation of these events as a way to encourage myself to keep going for longer periods of time. It may help you as well.

Opposite-side phenomenon

I discovered that in icing, sometimes pain that had been solved in one area would suddenly show up on the opposite side. Or that correction of one side of my body, would result in correction of the opposite side as well. I call this the "opposite-side phenomenon."

What you don't feel can hurt you

In icing, I repeatedly discovered that I was injured in areas that I didn't even feel. My years of icing experience have shown that there is often hidden pain, buried too deep to be felt. Removal of one layer of pain may result in a "new" pain somewhere else, that has finally come to the surface. That pain *was* there and *was* affecting one's physical condition; but only now is it available to be contacted and corrected.

Mild exercises

Incorporating a few easy twisting and stretching exercises into your ice therapy regimen reaps huge rewards. After you've reduced the inflammation and swelling of the connective tissue, joints and tissue that were out of alignment will fall into place with ease while you perform these low-impact exercises.

For me, yoga and icing together produced some incredible results. Although I attribute most of the eradication of my pain to ice therapy, there is no doubt that yoga helped to stretch and relax my muscles and had a calming effect on my mind as well. The wonderful part about yoga is that it's a low-impact exercise that doesn't irritate the tissue by heating it up.

Body putting you to sleep

In icing there is a phenomenon where your body suddenly puts you to sleep when it's ready to "back out" of a painful injury and needs the extra relaxation and non-intervention (from a conscious you) in order to let go of the pain. You fall asleep because, in the "backing out" process you are passing through a pain barrier that has to be breached so you can get to the next level. Sometimes on awakening you may feel pain but I've found that it dissipates quickly and that when it's gone there has been a marked change for the better, like improved range of motion or a chronic pain disappearing totally.

Letting-go reaction

Sometimes the day after icing you can wake up feeling horrible: achy muscles, exhaustion, "just feeling bad," and even disorientation. These unpleasant symptoms can even last for a couple of days, but they do go away. I found that it was my body's reaction to letting go of pain it had held onto for so long.

Section II: Where's the Pain

"I have had problems with my sciatic nerve for years - something with the nerve getting pinched, etc. It comes and goes but about 5 years ago it flared and between the chiropractor and Motrin, etc. I got worse every day. I was literally to the point that I could barely walk, drive, sleep, etc. In order to sleep I had to put a pillow between my legs and sleep on my side... You had always talked about icing and I finally was in so much pain that I did start icing on a more regular basis and for longer periods of time. I started icing whenever I even started to feel any spasm or a hint of a problem ... I swear by icing, at least for me."

— Brad Schutts, Palm Springs, CA

"I iced my right forearm for pain from tennis elbow. The first time I iced for 25 minutes, following with the twisting, stretching, and exercising. My right shoulder did "pop" during the exercising. I had no more pain for several days.

Later, the pain returned and I realized I needed to ice for at least 45 minutes or longer, so I watched TV for an hour and iced with the ice bag for a full hour, with the bag on my forearm again. After this session, I felt stiffness in the hand and wrist. I followed with stretching fingers, hand and wrist. I also noticed some tingling in the left side of the neck and top of the lower left leg, after about 45 minutes of icing. I did stretching exercises with no popping but the pain and stiffness in the hand, wrist, and forearm was gone."

— Ruth N., California

In the following section I'm telling you about my own pain because much of it may be similar to yours. It will show you how you can do what I did. Choose the chapter below that covers the part of your body where your worst pain lives. Read about what I did to remove mine by icing it. Then, don't delay: go right to Section III "The Method," and try it out.

Chapter 1: Head and Scalp

Scalp pain
Headaches/post-nasal drip
Temperomandibular joint pain (pain in the jawbone in front of the ear, also called the "TMJ")
Ears "ringing" and minor (trauma-induced) hearing loss

I was never bothered much with aspirin-level headaches, but after I began icing my neck and upper back, I started waking up with excruciating headaches that lasted five or ten minutes. My skull felt like someone had put a hatchet in it. After several days of waking up with this pain, I felt the top of my head and discovered that the pain was coming from the outside of my scalp, not the inside of my head. The scalp was sore along the suture line that runs down the center of the skull (Figure 1).

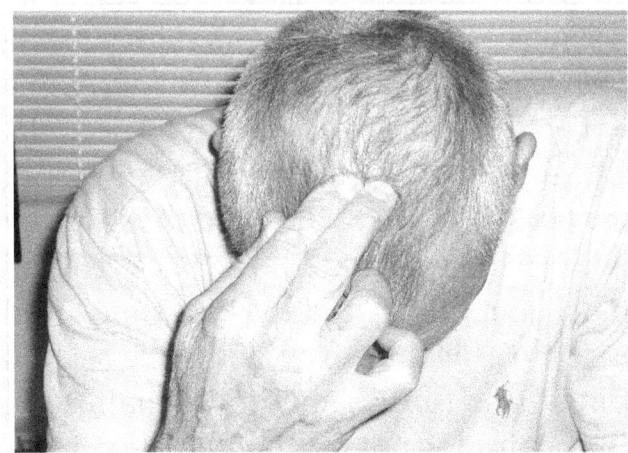
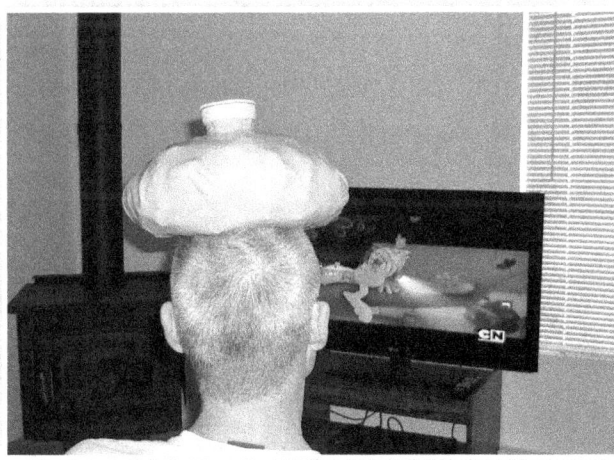

(Figure 1)   (Figure 2)

Over the next few months, I began working on my scalp. I iced the top of my head for one hour (Figure 2) and it was painful the entire time, but not that I couldn't stand it. The headaches stopped and the scalp was much less tender.

A few weeks later, the headaches started up again. When I put the ice on the second time, it hurt so badly that I was only able to tolerate the pain for forty-five minutes. During that time, I became dizzy and disoriented, and felt as if I wanted to throw up. Again, the headaches stopped and the scalp became even less tender. A few days later, I iced it again and was able to keep the bag on for two hours. At one and a half hours, the muscles in the right side of my face moved around for a few seconds like there were worms under the skin and then relaxed. When I did my twisting exercises, my upper back on the left side gave way with a loud crunch.

For many years, when I was eating (usually salads), if I got too big a piece in my mouth and tried to move it around with my tongue, an area on the right underside of my tongue would go into spasms and my tongue would freeze up. It was very painful and took several minutes to relax; for several days afterwards, the underside of my tongue was sore. Once, after icing the top of my head for three hours, when I did my twisting/stretching exercises, my upper back moved with a loud crunch. A few minutes later, I yawned and my tongue seized up...only this time the pain radiated deep into my right ear, up the side of my face to my cheekbone, the side of my nose, and up to the right side of my forehead. The entire episode only lasted a few seconds but it was shocking. I was completely taken by surprise. The energy being released was so intense, it left me open-mouthed with amazement. "Omigosh!"

I was certain that this was a turning point: a final correction or realigning of the connective tissue that was causing the problem. Indeed it was, because my tongue has never gone into spasm again. It's incredible to me that an injury to the top of my head was the cause of spasms in my tongue—and even more incredible was that exposing my skull to three hours of cold would fix it.

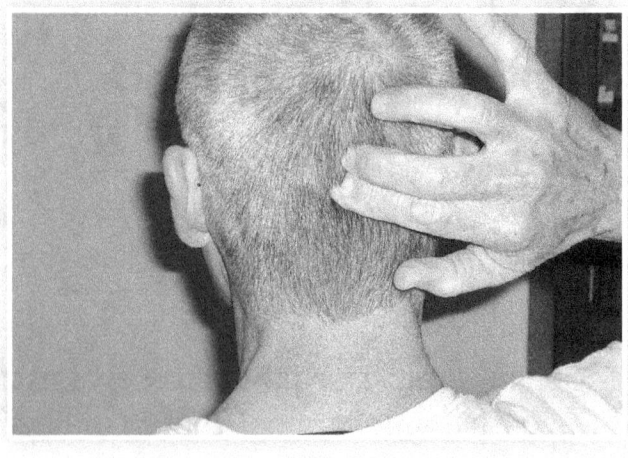

(Figure 3)

Next, I palpated areas at the back of my head (Figure 3) and iced those for an hour (Figures 4 and 5). The back reacted much like the top did the first time I iced it. It hurt for the entire hour. The second time I iced for an hour, it was painful in the beginning but the pain slowly subsided and I left that area as fixed. However, when I started working on my pectus excavatum (undue depression of the sternum or breastbone; see Chapter 5) it sensitized the connective tissue in both my neck and head, and I had to revisit those areas again.

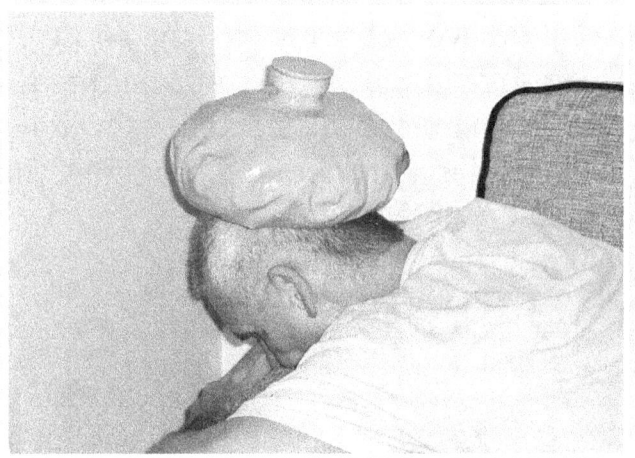 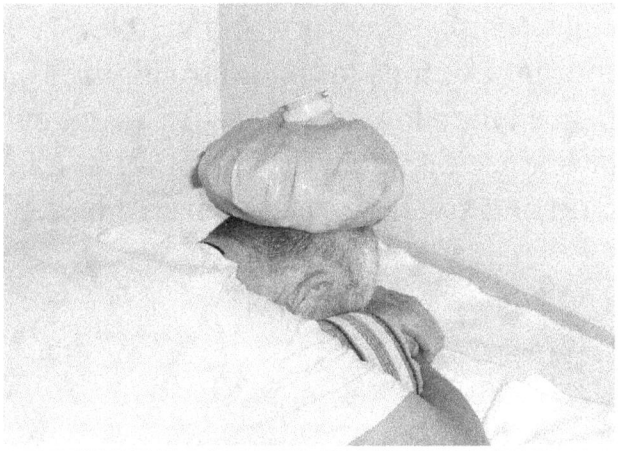

(Figure 4)   (Figure 5)

There are many muscles that ascend from the neck and attach to the skull. When I palpated the areas above my ears, they were very painful (Figure 6). The right side hurt more than the left, probably because of the weight of the camera and gear head pressing on my right shoulder and the side of my head and neck for the years I carried cameras. I iced both sides (Figure 7) for an hour each and felt the typical reaction of burning around the forty-five-minute mark. A few weeks later, I palpated again and found that they were still mildly painful, so I iced each side again for an hour. The icing was relatively painless and those areas were pain-free until icing my upper back sensitized them again.

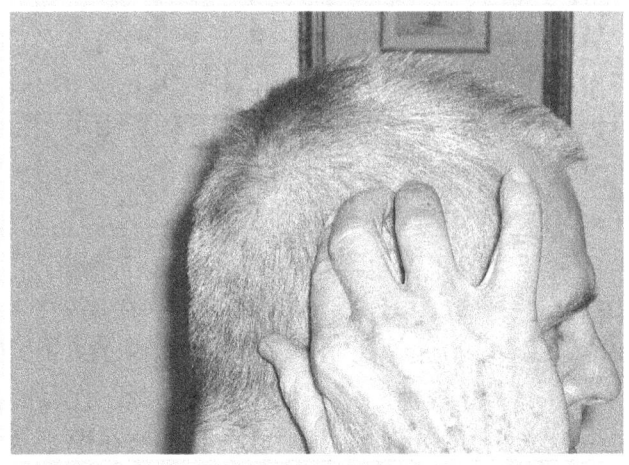 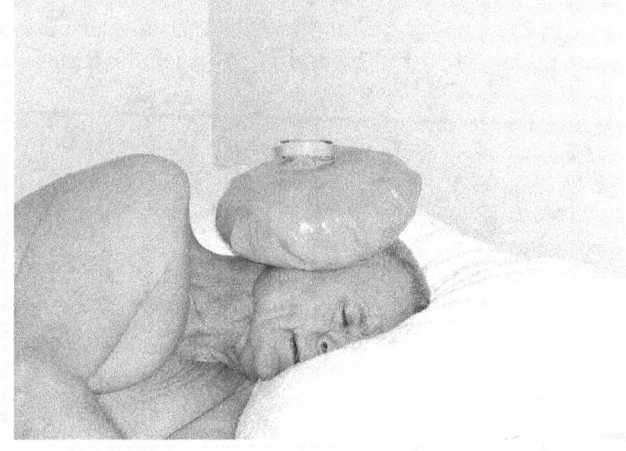

(Figure 6)   (Figure 7)

I recently took a trip that was very stressful and when I arrived, I was struck with shooting pain in my right temple and right eye. I was unable to do anything but hold my head in my hands. I began massaging my temples and found that the right temple was sore and swollen (Figure 8). I travel with my ice bag, so I iced the right temple for an hour (Figure 9). At forty-five minutes, the same pain reappeared for a few seconds with a vengeance and then quickly dissipated. A few days later, I iced it again for an hour and a half and besides a little mild discomfort, the pain and swelling disappeared. I haven't had that problem again.

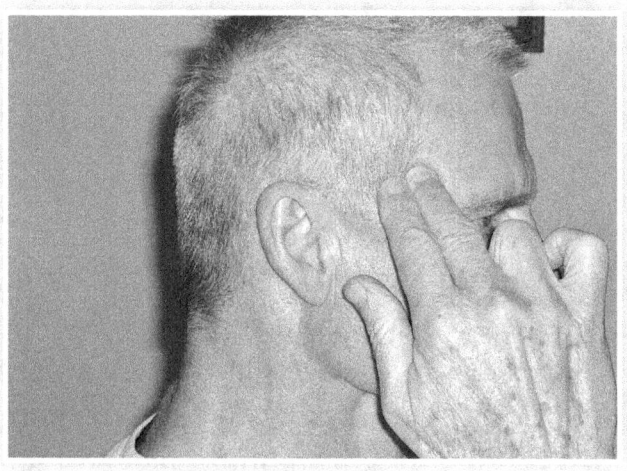

(Figure 8)

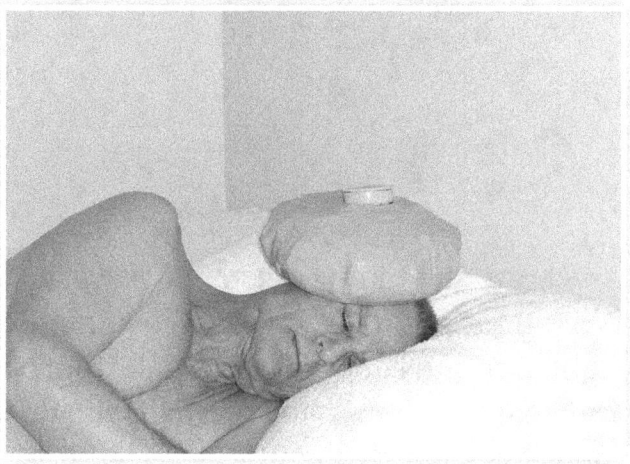

(Figure 9)

WHAT YOU DON'T FEEL CAN HURT YOU

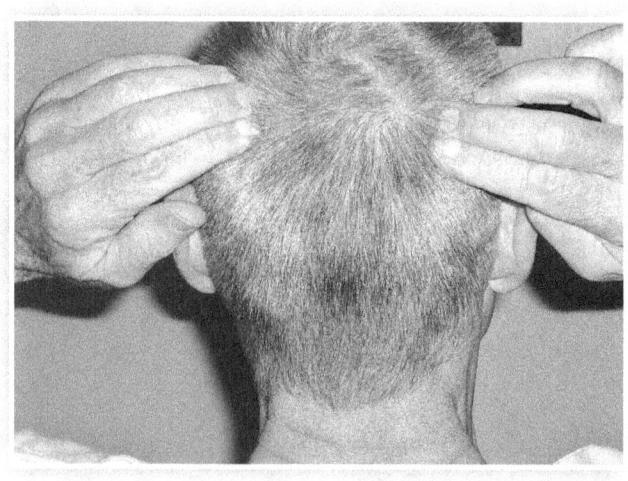

(Figure 10)

Along with the discovery of the time element of icing, I also discovered that I was injured in areas I didn't even feel. I had been icing the middle of my back and got a lot of movement in my thoracic spine (the 12 vertebrae of the upper back) after doing the twisting/bending exercises. In the third hour, I felt a sharp pain that radiated upward from my left ribs into my left pectoral muscle. The next morning, the pain was still in the pectoral muscle so I iced it for two hours. During the second hour, a pain began in my left upper back and radiated up my neck to both sides of the back of my head (Figure 10). These two spots came awake and were extremely sensitive.

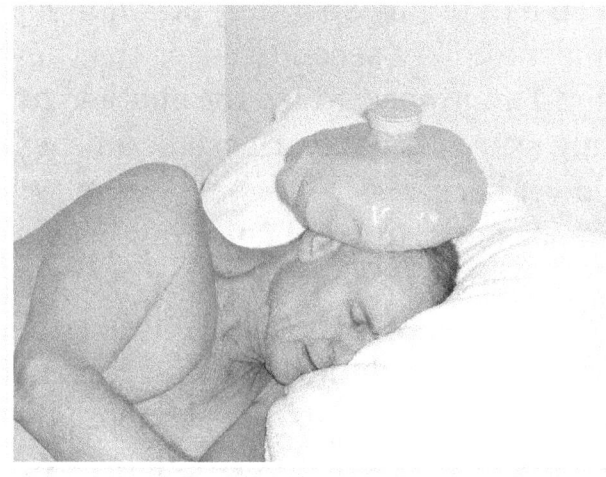

(Figure 11)

The following day, I iced each side (Figure 11) for one hour. I iced the right side first and was surprised that during the icing, there was more pain above the ear than at the tender spot. After the hour, I did my twisting exercise and felt and heard a loud crunch in my upper back on the right side near the scapula, or shoulder blade. I iced the left side next and the sore spot was painful for about forty-five minutes into the icing. After that, the pain diminished and at the end of the hour, when I did my twisting exercises, there was no movement in my upper back.

One morning I was icing the right side of my neck (Figure 5 in the chapter on the cervical spine) and noticed that the left side of my forehead above the eye was becoming extremely painful. After I finished icing the neck, I palpated the painful area and discovered that both areas above the brow were tender (Figure 12). I centered the ice bag on my forehead and frontal sinus (Figure 13) and was in for quite a shock. In the center of my forehead between the eyebrows began a pain so vicious that my stomach turned and I thought I was going to throw up. The pain intensified and my abdominal area continued to churn for three or four more minutes.

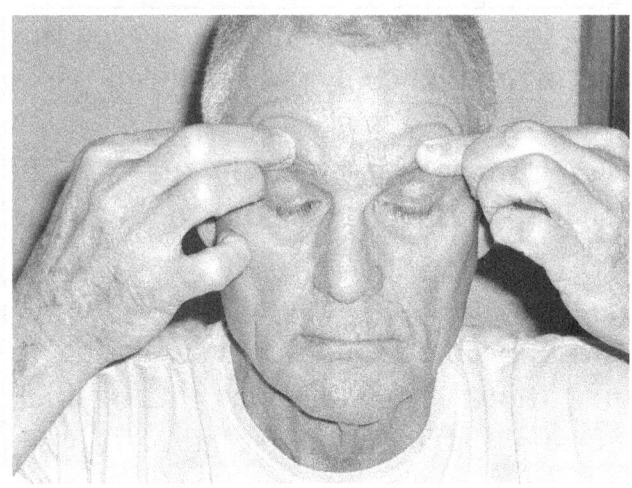

(Figure 12)

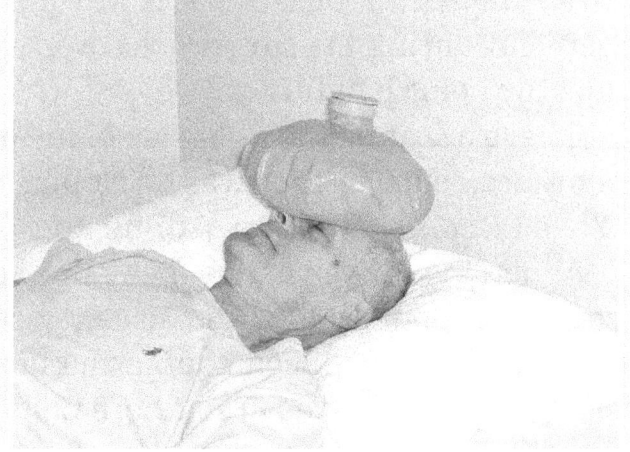

(Figure 13)

I didn't think I was going to be able to hang on, but then my stomach started to settle down and the pain in my head began to diminish. At about fifteen minutes, a feeling like electrical sparks (I pictured a Fourth of July sparkler) began in my lower abdominal/bladder area and lasted maybe thirty seconds. After that, my forehead began to feel more comfortable and I continued to ice for another fifteen minutes. When I was doing my stretching exercises, I experienced crunches, cracks, and pops in my neck and upper and lower back.

The next day, I iced again, this time for an hour. In the beginning, it was painful but not nearly as bad as the first time, and there was no nausea. At the magic moment, after forty-five minutes, the area between my eyebrows became very painful for several minutes. I had a light-bulb moment and remembered that in the auto accident that broke my left humerus (the bone of the upper arm), my nose was also broken and a big gash opened between my eyebrows. I decided that those old injuries (and possibly the frontal sinus) were probably responsible for the intense pain. During my stretching exercises, I heard a loud crack in my lower back.

When I put the ice bag on my forehead, it also rested on the bridge of my nose. During an icing a few weeks later, pain in my nose became so intense that it felt like it was broken and needed to be put back in place. I moved my nose back and forth with my fingers and didn't seem to get much movement, but afterwards the pain slowly went away. It reminded me that seven years previously, I'd had a basal cell carcinoma removed from the bridge of my nose. At that time the doctor drew a circle about the size of a dime on the bridge of my nose with a felt marker and proceeded to cut it out. I declined a skin graft, so the doctor closed the wound with three staples. I realized that it was probably the scar tissue and the nerves that were cut during the surgery that were reacting to the cold, as all injured tissue does, by being painful.

I have a scar on the right side of my neck below the jaw that is the result of one of my accidents; an area was torn open and a lot of damage was done to the tissue. When I iced it, it was very painful, too.

When I was five years old, my parents had my tonsils and adenoids removed, and throughout my life I had no ear/nose/throat problems whatsoever. During my forties, my sinuses began bothering me and I thought maybe I was getting allergies. I had always lived in Southern California and I attributed it to dirty air. When I was fifty-six, I contracted a vicious head cold coupled with a bad case of bronchitis.

The doctor put me on a two-week antibiotic regimen and, at the end, I was much better but continued coughing up phlegm.

This continued for months so I went to the doctor and he told me that I had a chronic lung infection caused by post-nasal drip, and he prescribed anti-histamines. I took them for a few weeks but couldn't handle the side effects of dry mouth and fatigue so I stopped and the symptoms returned. Since I felt the cure was worse than the disease, I decided to live with it.

Five years later, I was lying on the couch, watching TV and icing the front of my right ribs just under the pectoral muscles. Suddenly I felt like I was going to sneeze, only the feeling was five times as intense. For several seconds, I had that itchy, ticklish feeling one gets before sneezing, but I didn't sneeze. Afterwards, my sinuses drained into the back of my throat for half an hour and I was spitting out gobs of mucus. When it was over, my sinuses felt like I could have driven a truck through them. Over a period of time, they filled again, but after that the post-nasal drip was less and I was coughing up less phlegm out of my lungs.

A few weeks later, I was icing the middle of my upper back when I began to sneeze uncontrollably. I sneezed at least six times with barely a breath in between, and afterwards my sinuses drained again; after that, my sinuses and lungs dried up a little more. As a result of icing painful areas on my head and neck, and fixing my pectus excavatum, the post-nasal drip stopped and my chronic lung infection went away.

In the mid-1970s, I was working on a TV series and the director fired a movie pistol with a full load in it next to my right ear; when the gun went off, it felt as if a knife were stuck in my ear. My shoulders shot up and my head and neck shrank into my body like a startled turtle. I couldn't hear for a few minutes and when the hearing returned, my right ear was ringing; it never stopped until I began icing my skull.

Looking back on my therapy, there isn't one area that I iced that didn't cause a reaction in my head. My right eye ached; I had sharp, stabbing pain in my right ear, pressure and pain in my sinuses, and times when my entire head felt like it was being put in a vise. During the icing process, I also became aware that not only did my right ear ring, but there was a harmonic noise in my left ear as well. One afternoon, while icing the flat spot behind my left ear (Figures 14 and 15), in the third hour the harmonic noise stopped. The right side was not that simple, but by icing the right side of my head above, behind, and below the right ear, the noise slowly disappeared.

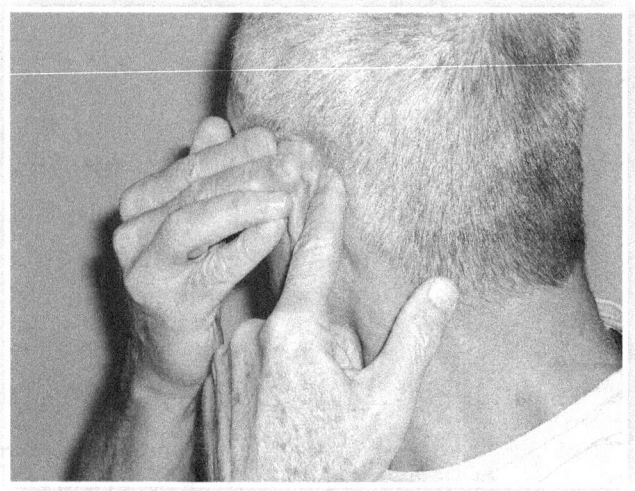

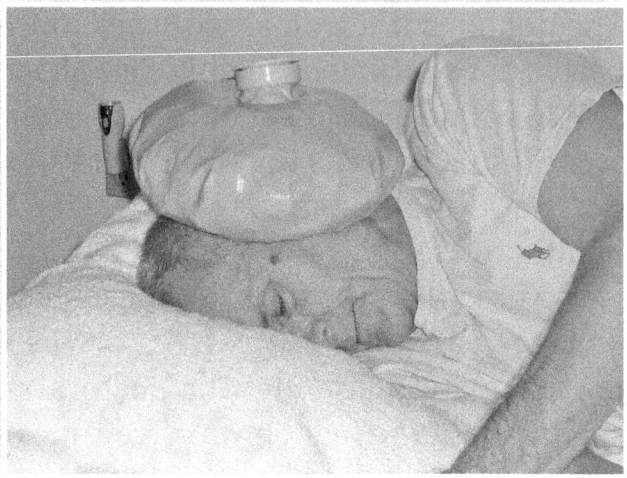

(Figure 14)

(Figure 15)

For years, I was bothered with popping and pain in the right temporomandibu-lar joint (the jawbone joint in front of the ear, also called the "TMJ"). I palpated the jaw muscle and the depression below the ear (Figure 16), and both were very painful. I iced it by centering the ice bag here (Figure 17). In the third hour, the jaw muscle became painful and then it untwisted and relaxed, and the pain went away. There was also pain at the zygomatic arch (cheek bone) and a sharp pain in my ear that subsided in the third hour as well. As the result of several icings, I no longer have pain or popping in the right TMJ.

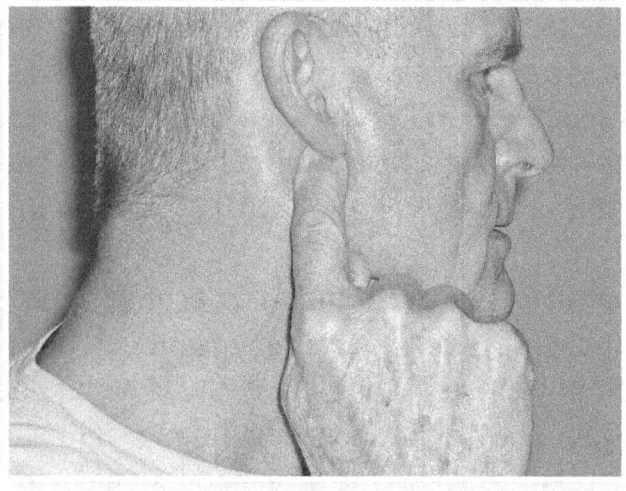

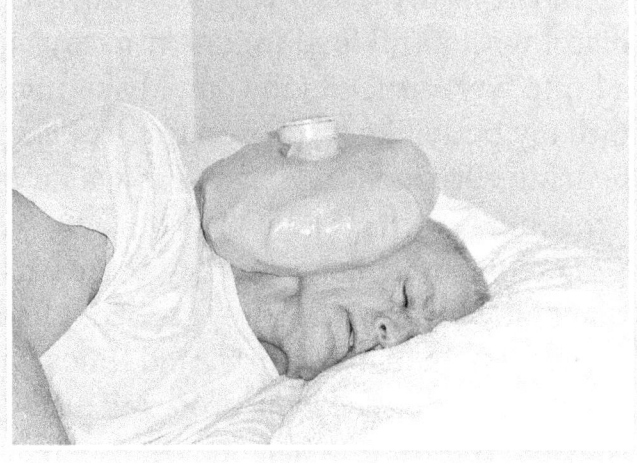

(Figure 16)

(Figure 17)

At the very end of the icing process, when I thought I had solved my head pain and "ringing" ears, this happened: I woke up one morning and the flat spot on my skull behind my left ear was burning. I put the large ice bag on it and, during four hours, experienced shooting pain in my right ear (that "opposite-side phenome-non" that I have experienced while icing), the muscles in my left jaw and left side of my neck releasing, pain in both eyes, and reactions in other parts of my body. When I did my stretching exercises, I felt and heard alignments in my neck, shoulders, and upper and lower back.

But the next morning when I woke up, the harmonic noise in my left ear had returned with a vengeance. I got up, made coffee, and poured my juice, then turned on the TV to watch the morning news. When there was any music, it sounded like I had a tiny piccolo player in my ear loudly echoing the sound. I tried not to panic but since I was in Costa Rica and wasn't returning to the U.S. for three months, I knew I had to deal with it myself. I continued to ice all around the left and right ears and the left side of my neck, and over three weeks, the echo subsided.

For years, I had experienced a mild hearing loss in my left ear caused by trauma to my head. When I turned my right ear to a sound, it was loud and full, but when I turned my left ear to the sound, it was less loud and full; however, since I wasn't constantly having to say, "What?" when speaking to people, I was having no problem living with it. Incredibly, after the echo went away, the hearing in my left ear matched the hearing in the right.

My take on what happened is that when I iced that first time behind my left ear, it caused fluid to gather in the ear and the fluid is what caused the distortion of the sound; when I tapped on the side of my head near the ear, it sounded like slapping a ripe watermelon.

Chapter 2. Cervical Spine (vertebrae of the neck)

Neck pain (limited range of motion)
Migraine headaches

The damage to my neck was caused by two incidents. The first was a car accident I had in my early twenties when I got a whiplash injury that affected the left side of my neck; the second was due to the years of carrying cameras on my right shoulder, which injured the connective tissue and nerves on the right side of my neck and head. As with many of my injuries, the physical manifestations over the years were pain and decreased range of motion.

Early on when I began icing with gel packs, my neck wasn't an issue because I was icing lying on my back with my head on the packs, which was comfortable even though it wasn't very effective. When I started using the ice bag and was icing face-down on my stomach, I found it difficult to ice for any length of time because my neck was so stiff. So before I could really get into long sessions while lying on my stomach, I had to fix my neck.

When I iced my neck for the first time with the gel packs, at forty-five minutes I was jolted by an electric shock that shot up the right side of my neck and head, and down the right side of my upper back. A few days later, I was returning home from a trip to visit my family when suddenly I couldn't turn my head. My first reaction was panic because I was driving at sixty-five miles per hour. I forced my head to the right and then to the left, and each time I heard loud crunching sounds like rattling the bones of a skeleton.

Suddenly, all my senses seemed enhanced: the light was brighter, sounds were louder, and I felt hyper-aware and light-headed. I kept turning my head to the left and right and discovered that the range of motion had increased dramatically in both directions. This occurred in the early stages of my icing and I was astounded that such a dramatic adjustment could happen as the result of icing for such a short time. I was also surprised that the adjustment happened several days after the treatment.

DON'T GIVE UP!

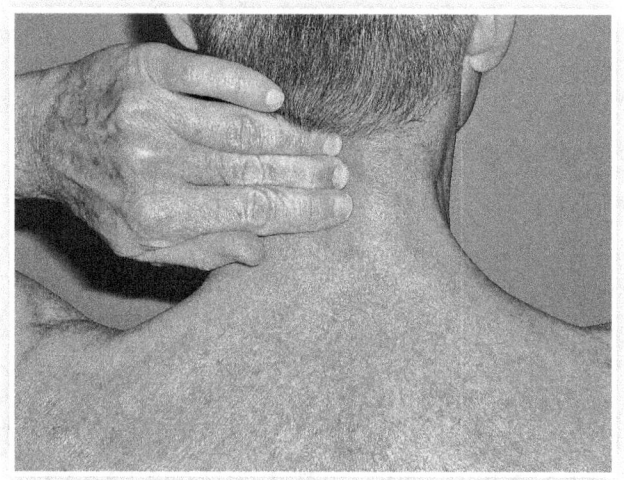

(Figure 1)

Icing the back of your neck is one of the most difficult positions to stay in for a long time, but I devised a way to make it tolerable and even comfortable if you stick with it. After you've palpated your neck for pain (Figure 1), fold two towels of the same size long ways twice, roll them up, and place them side by side (Figure2). Place your head between them and bring your arms up and around the towels to hold them in place (Figure 3). I've been able to ice my neck for four hours this way and even get comfortable enough to fall asleep. The first ten or fifteen minutes are the hardest, but if you hang in there, it becomes more comfortable.

(Figure 2)

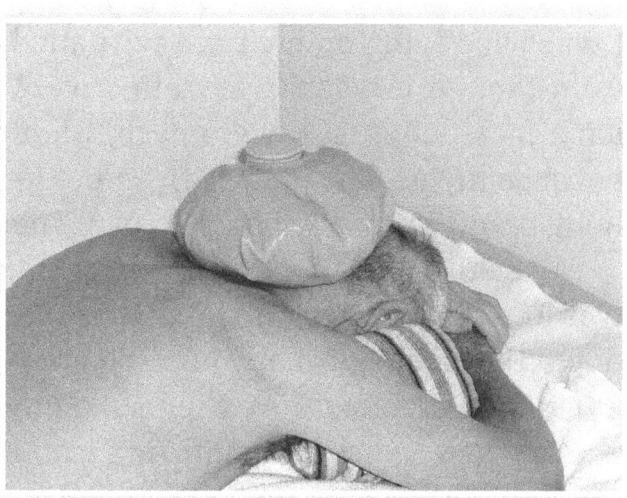

(Figure 3)

There are two areas of the spine and the surrounding tissue that are prone to injury, just because of the way we are built. These are the two junctions where lordosis (the forward curve of the neck and low back as viewed from the side) meets kyphosis (the backward curve of the upper back and buttock as viewed from the side). The last vertebra of the low back and the first vertebra of the tail bone (sacrum) is the junction that most people have trouble with, because it bears the weight of the entire upper body.

However, the junction of the last vertebra of the neck and the first vertebra of the upper back is another area that can be responsible for pain too, even though it only bears the weight of the head and neck. The stress at this junction can be partly responsible for pain in the muscles arising from the shoulders and running up the sides of the neck, as well as headaches, ringing in the ears, upper back pain and pain in the hands and fingers. Place the ice bag on the bump in your spine at the base of your neck (the junction of C-7, T-1).

During one of these sessions, a pain began that encircled my neck and head at the level of the ears. It felt as if a wide metal strap were being tightened around my head until I thought it would burst — and then it quickly subsided. That same crushing pain has happened in my upper and lower backs as well.

One of the more bizarre results occurred after icing the lower portion of my neck. After doing my stretching exercises, I happened to reach up with both hands and press on the back of my neck just above the seventh vertebra, which is the big bulge at the base of the neck. The only sensation I can compare it to is when I first tried a new candy introduced back in the 1970s called "Pop Rocks." They exploded like popcorn in my mouth and made loud popping sounds. The connective tissue in my neck did the same thing. When I was pressing and massaging that area, there were loud popping sounds and it felt like the tissue below the skin was crackling and crunching. I kept massaging the area and, after a few minutes, the crunching finally stopped. A few weeks later, I iced the same area and the same thing happened but to a much lesser extent. Eventually, it went away altogether.

When I started icing my neck with the ice bag for long periods of time, the connective tissue and joints in my neck relaxed and realigned themselves with a little help from my stretching and bending exercises during and after icing. Not only did my neck get better, but the icing also helped free up areas in my ribs, shoulders, and upper and lower back. This process of recovery through ice therapy has taught me that "everything is connected."

After working on the back of my neck, it became apparent when I palpated the sides (Figure 4) that there was a lot of damage there, too: on the left from the whiplash and on the right from carrying cameras. I began icing on and off (between working on other areas), with incredible results (Figure 5). Also, the muscles arising from my shoulders were incredibly tender (figure 6), so I iced them as well (figure 7). It's impossible to try to relate every pain, pop, and crunch that occurred as a result of icing the sides of my neck, but I did feel a lot of pain in my eyes and ears, and pressure in my sinuses. I'm sure icing my neck contributed to the cure of my post-nasal drip and the ringing in my ears.

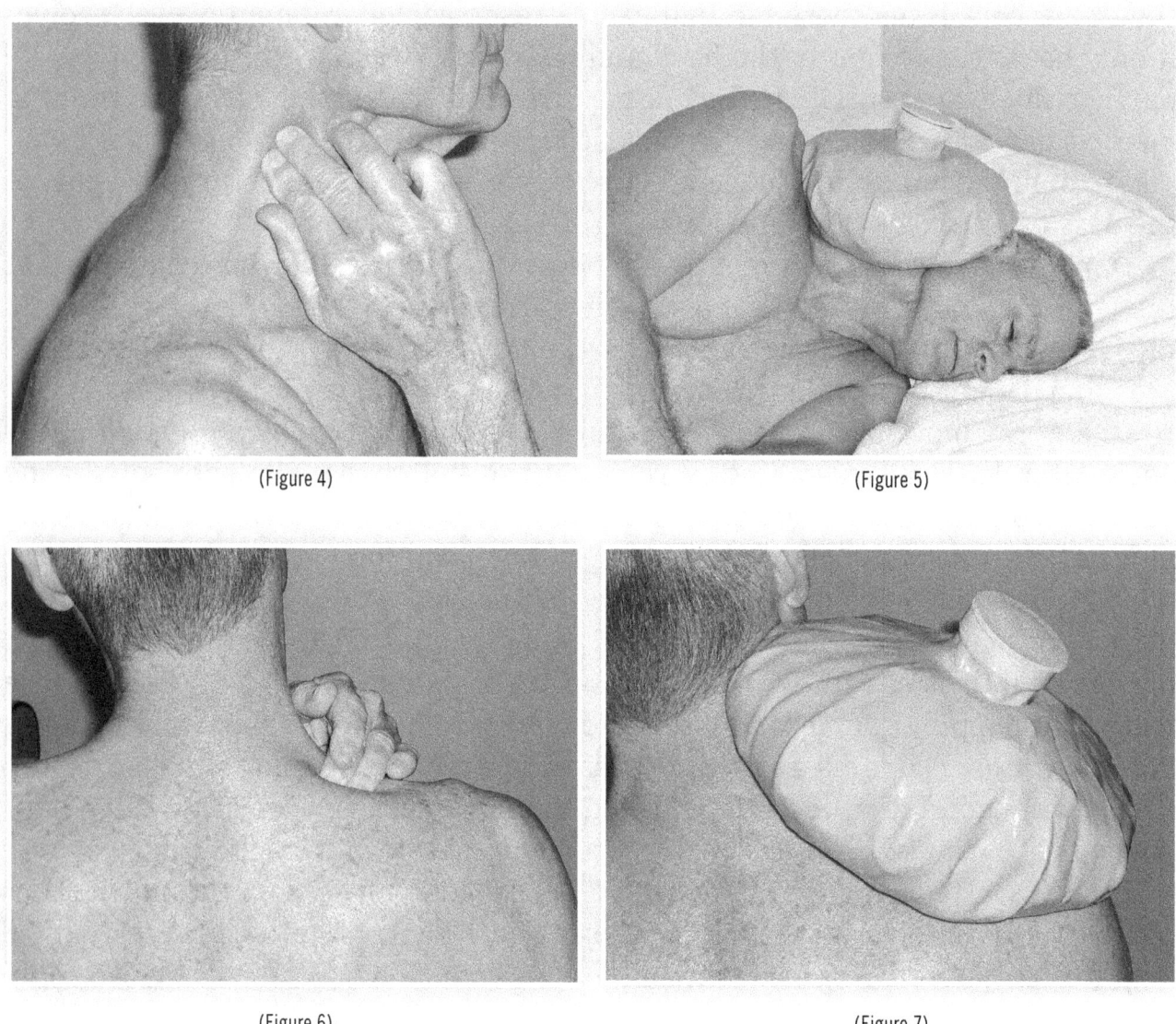

(Figure 4)

(Figure 5)

(Figure 6)

(Figure 7)

I want to tell you about a major event that happened on the right side. On one occasion, during the second hour — as often happens — my body put me to sleep. I awoke with what felt like a three-hundred-pound gorilla sitting on my head and neck. The ice bag felt like it weighed a ton but I resisted the urge to throw it off. Since I had experienced huge build-ups of pain during previous icings, I understood that I was "backing out" of an old injury and was passing through a "pain barrier" that had to be breached in order to get to the next level. The crushing pain only lasted fifteen seconds or so and turned out to be a major breakthrough on the road to fixing that side of my neck. That breakthrough peeled away three or four layers of the onion.

Another breakthrough happened on the left side after I fell asleep again during icing. When I woke up, I had the mother of all headaches. My forehead, eyes, and temples were exploding. My first instinct was to get up but then I realized that, like so many other times when I'd had these huge increases in pain, it would dissipate quickly. During my stretching/bending exercises, that side let out a huge *thunk*; the pain decreased and the range of motion increased.

The next area of pain that "woke up" because of icing all around it was the front of my neck on either side of my Adam's apple. I found pain there (Figure 8) and iced it as shown in (Figure 9). For some reason, these areas were particularly painful in the beginning, especially an old scar on the right side. Icing the front hurt so badly that my entire neck and head throbbed. Each time I iced, the pain became less and less until finally it was gone. The front of my neck is another example of "hidden pain" that I had been living with for decades; since everything is connected, I'm sure those injuries contributed to my overall condition of unwellness.

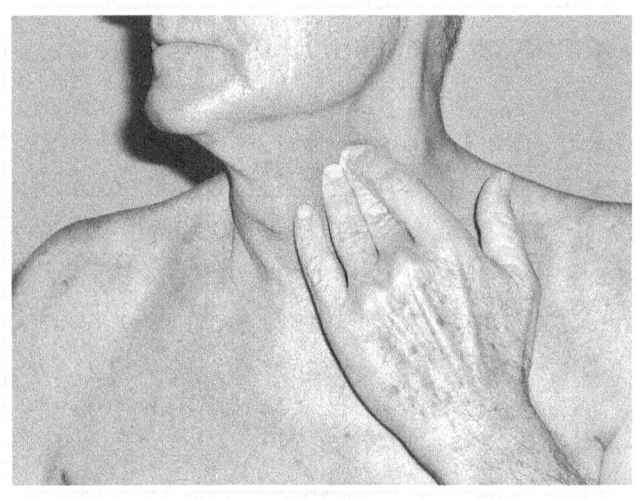
(Figure 8)

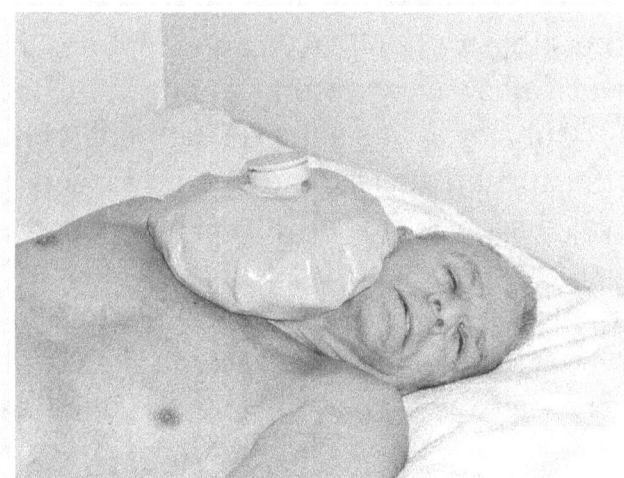
(Figure 9)

Later, while I was working on the pain in the connective tissue and joints of my hands (see Chapter 4), because there are these connections, I should not have been surprised that when I bent my head to the right and left, my neck cracked and *thunked* and my head dropped farther and farther toward my shoulders. When I leaned my head back, the same thing happened and the range of motion increased dramatically in that direction as well. It's hard to believe that icing the hands causes a release in the neck, but if you look at a nerve root chart, it makes sense because the nerves in the arms and hands exit the vertebrae in the neck.

In my early twenties, I began experiencing visual irregularities when driving at night. The bright lights triggered a partial blindness that would force me to pull over to the side of the road and wait until my vision returned to normal. After the visual impairment, I would get a violent headache that sometimes made me vomit; early on, I learned that the longer the "blindness" lasted, the more severe the headache and that being around bright lights would trigger it.

A doctor recommended an electroencephalogram (a recording of currents emanating from nerve cells in the brain) and the results uncovered a spike in my brain waves during periods of light sleep. He diagnosed me as having petite mal epilepsy and prescribed Phenobarbital. At the time, I was working at a dangerous job and had to choose between it and the medication. Since more episodes had occurred while I was taking the medication, I quit it and kept the job. As time passed, I learned how to control the seizures by not allowing things to "get to me" and trying to remain stress free. When I did get an attack, I tried not to panic and allow anxiety to make it worse. Many years later, I was talking with an acquaintance and the subject of epilepsy came up; I described my symptoms and he said that those were also the symptoms of migraine headaches.

One day, I was icing the left side of my neck and a couple of hours later, I had a "petite mal" attack. The next day, I iced the right side and I had another attack. This was disconcerting because I hadn't had an episode in quite a while and I wasn't under any kind of stress. I became even more suspicious because the nature of these spells was different from any I'd had in the past. When they were over, I felt like a ten-pound weight had been taken off my shoulders and as if a layer of pain was stripped away. I had the feeling I had been living with a low-grade headache all my life.

A thought came to mind, a connection I'd never made before. The first episodes happened a year or so after my "whiplash" auto accident in my early twenties. I'm convinced now that I was misdiagnosed and that it was the injury to the left side of my neck that caused the "epilepsy." I've never had an episode since. If you've had a neck injury (that you're aware of) and you're experiencing similar symptoms, try icing your neck.

Chapter 3. Shoulders/Clavicle (collarbone)/Upper Arms

Shoulder and upper arm pain
Shoulders and clavicle "cracking"

During my lifetime, my shoulders sustained injuries piled on top of injuries. They resulted in pain, popping and cracking, and numbness in my left hand when I slept on my left side. My left shoulder and upper arm were trashed during a car accident that broke my left humerus. My right shoulder and upper arm were trashed carrying cameras and hauling around heavy camera equipment. I don't know how I managed it, but I was lucky enough not to have torn a rotator cuff. (That's the structure the joint of the upper arm rotates in.)

When I worked for an orthopedic group, I consulted the shoulder specialist about the pain in my left shoulder and numbness in my left hand. He took X-rays and told me that the humerus had been set incorrectly (twice—the first time while I was awake in intensive care and a few days later in surgery), and that the acromial process (part of the scapula) was pinching a nerve when I slept with my arm up over my head. He suggested arthroscopic surgery (using an endoscope for carrying out diagnostic and therapeutic procedures within the joint) to grind down the underside of the acromial process, giving more room between it and the head of the humerus. I had the surgery but it didn't solve the problem.

When I first began icing, I ran into the same problem with my shoulders that I'd had with my neck. When I was icing on my side, the shoulder I was lying on would become so painful after an hour or so that I would have to stop. And when I was icing the back of my neck with my arms up over my head, the fourth and fifth fingers of my left hand would go numb and I'd have to stop that, too. So in order for me to continue trying to fix the rest of my body, I had to ice my shoulders back to health first.

One thing about our skeletal structure is that our shoulders are basically stuck out in mid-air. The only attachments to other bony structures, and the only things that keep the shoulder girdle from collapsing into the ribs are the clavicles (collarbone), scapula (shoulder blade) and, of course, the connective tissue that surrounds it.

Since there are so many areas of pain connected with the shoulder girdle and clavicle, I'm going to list them all together and then discuss the results. Palpate and, if painful, ice these areas: Figures 1 and 2, Figures 3, 4 and 5, Figures 6 and 7, Figures 8 and 9, Figures 10 and 11, and Figures 12 and 13.

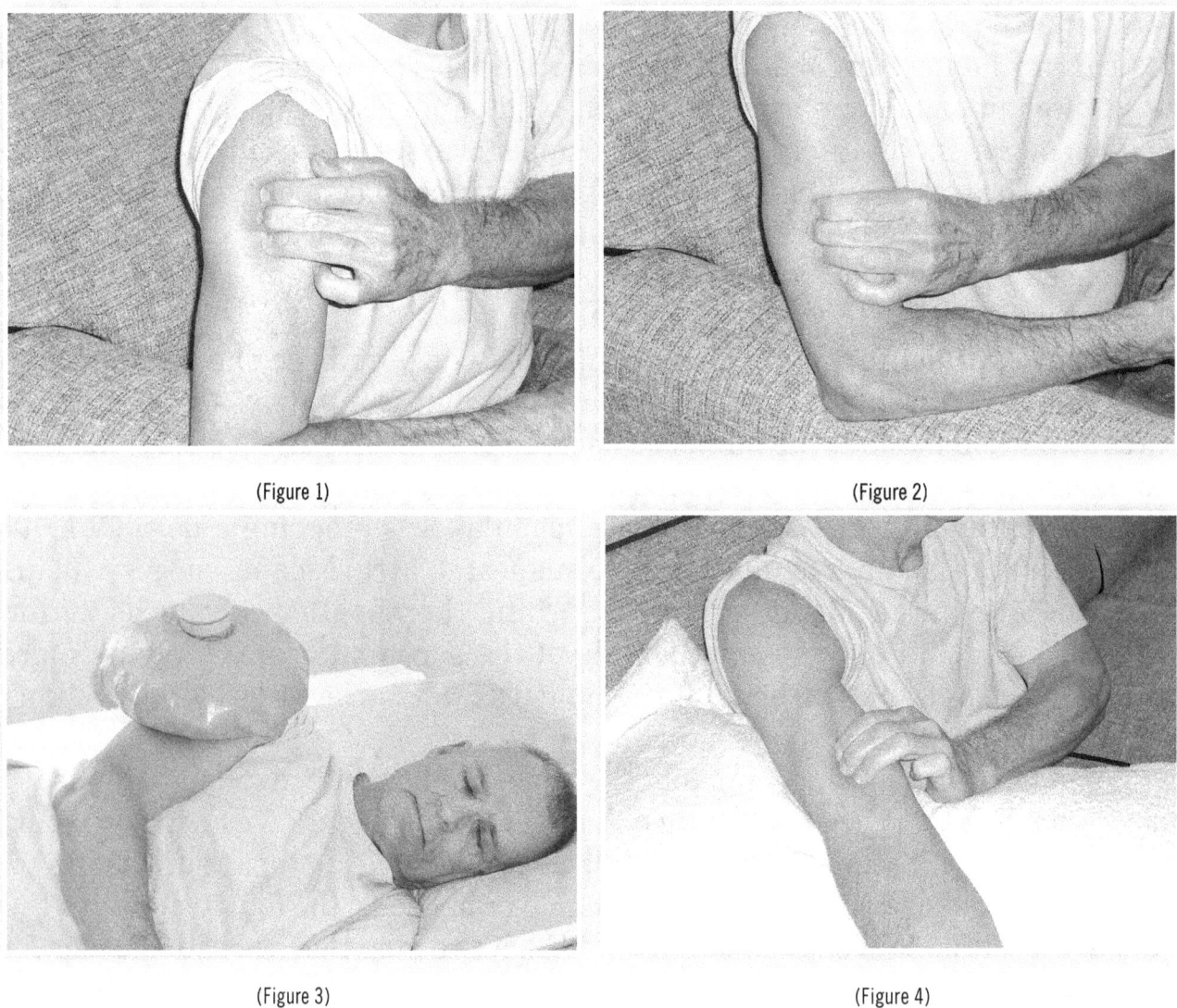

(Figure 1)

(Figure 2)

(Figure 3)

(Figure 4)

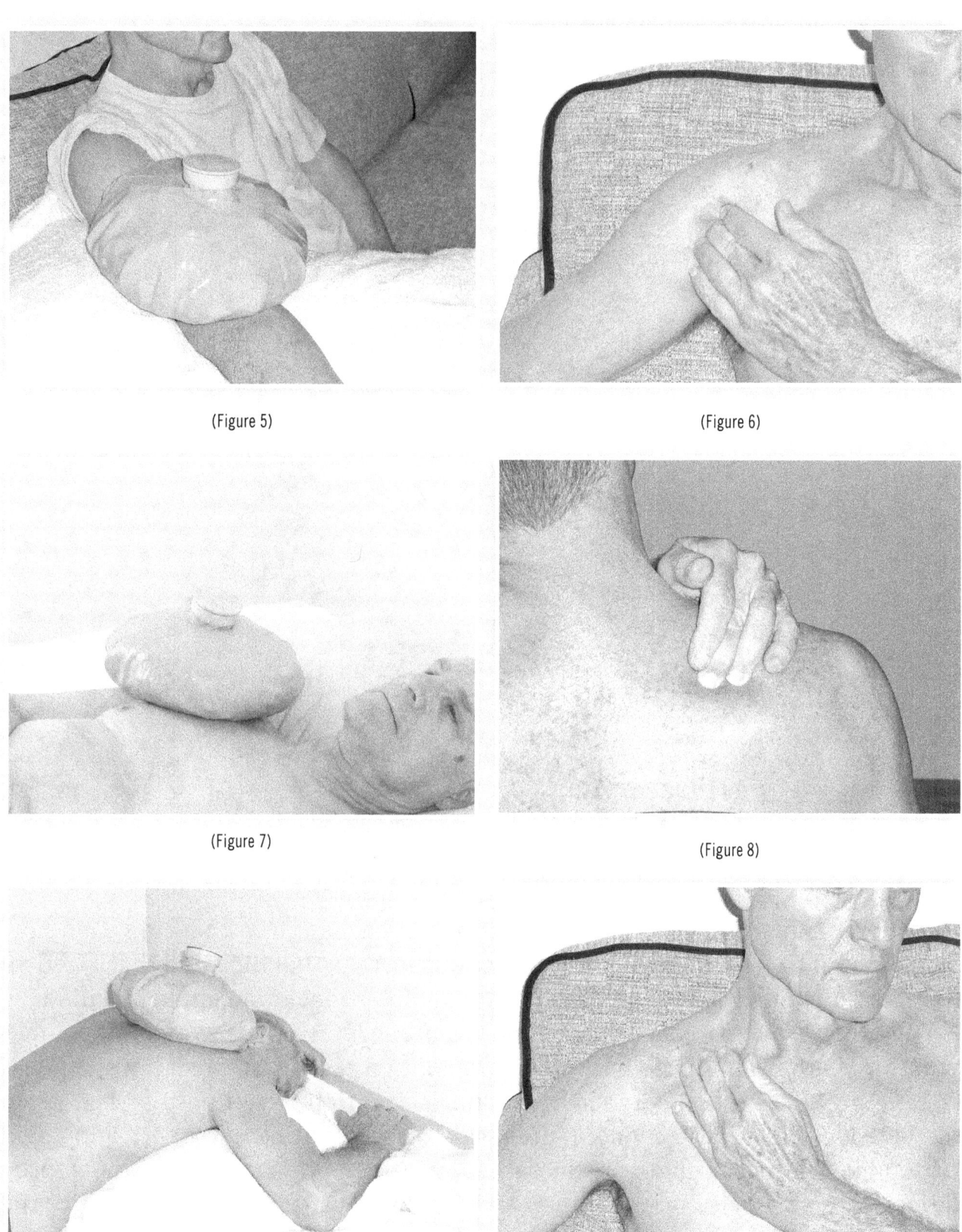

(Figure 5)

(Figure 6)

(Figure 7)

(Figure 8)

(Figure 9)

(Figure 10)

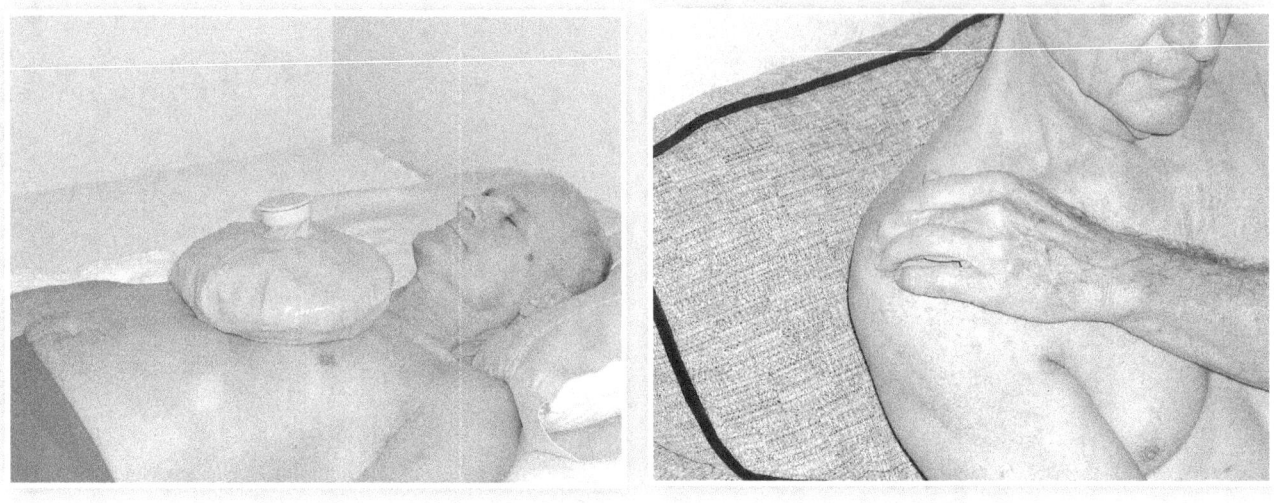

(Figure 11)   (Figure 12)

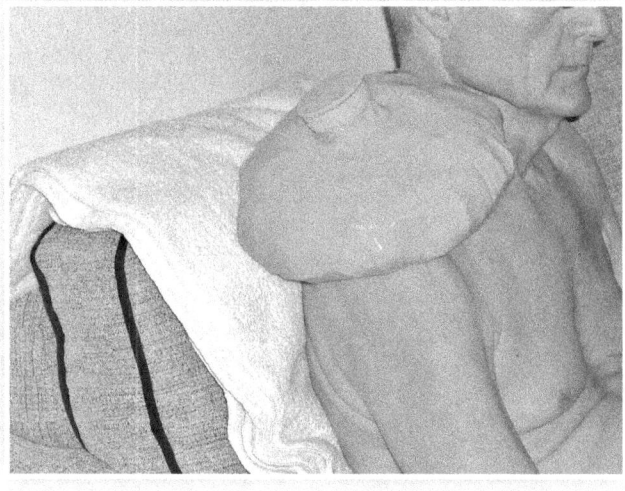

(Figure 13)

Rather than try to describe the reactions of each area to the icings and how long I iced each, I'll just say that the reactions were astounding and that I iced all the areas up to four hours. Here are a few examples that stand out in my mind.

My clavicles were injured on both sides, but the left was the worst. I remember that, for years, I could rest both my hands flat on a counter, push my shoulders up, and my clavicles would crack and crunch. The first time I put the ice bag on the left side (Figure 11), my neck and head began vibrating with a crushing pain so debilitating that I could only tolerate it for fifteen minutes. After two or three times, I was able to ice for more than an hour and eventually I was able to ice the pain away.

The lateral end of the clavicle at the top of my shoulder was painful, too (Figure 12). The first time I put the ice on it (Figure 13), at forty-five minutes I felt a "lightning streak" of pain up the side of my head and down my arm to my hand. After each icing, I always performed my twisting and bending exercises; my shoulders and clavicles moved around with cracks and crunches, and I could feel the entire shoulder girdle shift.

After I had dealt with the most painful parts on both shoulders, I was able to resume icing while lying on my side. As I've done with the rest of my body, I revisited painful areas whenever it was necessary. Now, my shoulders and clavicles are pain-free and I don't wake up in the middle of the night anymore because of pain in my left arm and numbness in my left hand. Icing the connective tissue in my neck and upper back also caused realignment in my shoulders.

As one thing led to another, I became aware that my right biceps was burning, so I palpated it (Figure 4) and discovered that the muscles were extremely painful. I began icing it while lying on the couch watching TV (Figure 5). The result was an achy, burning feeling that increased and decreased in intensity as the hours went by; then at one point I felt a muscle spasm release, and the result was a decrease in the pain I was experiencing in the muscles of my right upper chest.

As part of the waking-up process, I became aware of pain in both my biceps. I palpated them and they were both extremely painful to the touch. How could I live for decades with pain in my biceps and not know it? It's simple: I never bothered to check.

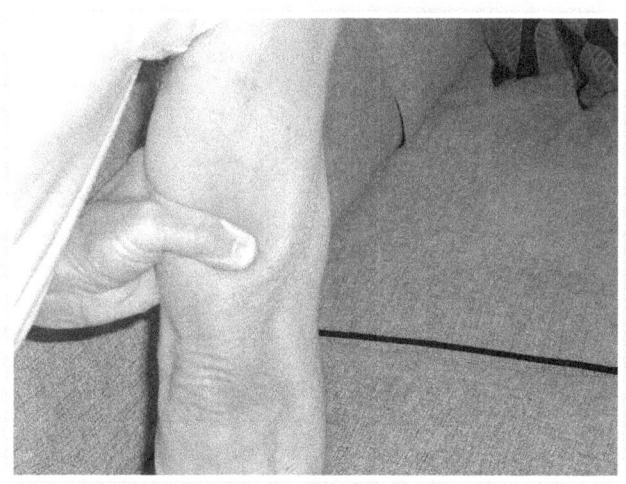

(Figure 14)

I figured that if my biceps were thrashed, my triceps probably were too, and sure enough, they were (Figure 14). When I worked as an assistant cameraman, my right arm was always slung up over the top of the tripod to keep it on my shoulder, and that kept the triceps in a constant state of extension, causing the injury.

I iced these muscles lying on my stomach as shown in (Figure 15), which is the most efficient, but awkward and difficult to do. Icing as shown in (Figure 16) is easier, and you can do it while watching TV. At one point, a lightning-like pain shot up the back of my shoulder, up the side of the neck to the bone called the mastoid process (the prominent bulge behind the ear), deep into my right ear, and up the side of my head. Icing my upper arms also caused movement in my ribs and scapula during my exercises, and helped resolve the tennis elbow in my right arm (see Chapter 4).

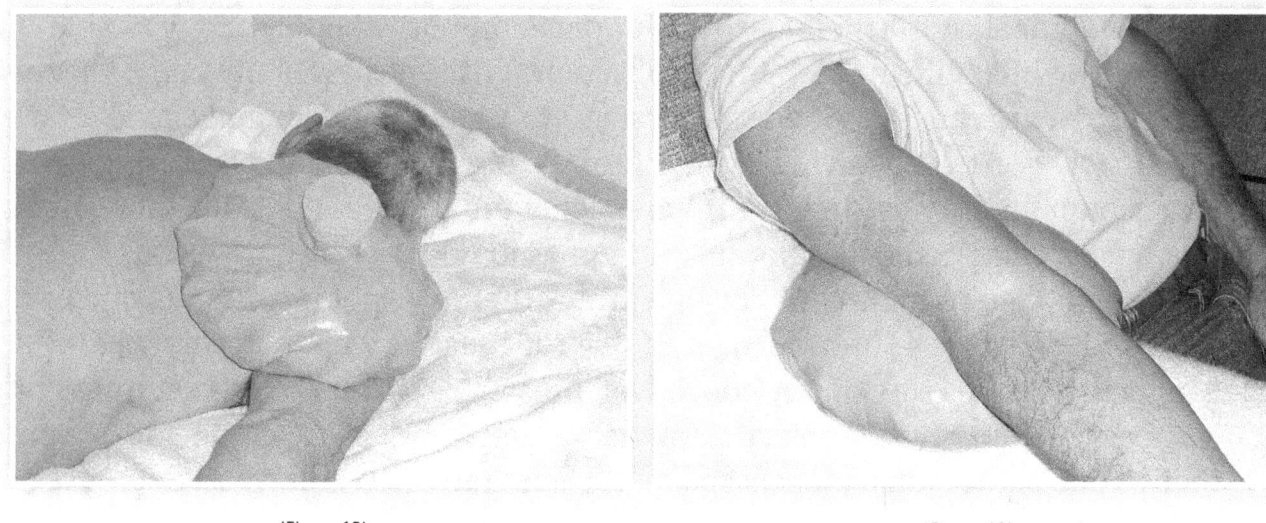

(Figure 15)   (Figure 16)

Chapter 4. Fingers/Hands/Wrists/Forearms/Elbows

Pain in the joints of the hands and fingers
Wrist and forearm pain (carpal tunnel syndrome)
Elbow pain (tennis elbow)

The years I worked as an assistant cameraman threw my entire body completely out of whack. A large part of the job was getting the camera where it needed to be, so I was constantly securing or unsecuring the cameras and gear heads from different-sized tripods, hi-hats, dollies, and cranes. One thing an assistant cameraman learns early on is not to cinch things down too tightly because if you do, it can be hard as hell trying to loosen them up later. I sprained my hands and wrists many times in that job, and I remember pain shooting up my arms, zapping the strength from my hands and wrists.

When I started working in health care, I did a lot of computer work, adding "repetitive motion" injuries to the mix. Then when I retired, I began baking my own bread. Kneading the dough uncovered the damage to my right hand and wrist and caused me to start icing them.

When I did begin icing my right wrist the pain was so intense I could barely leave the bag on for fifteen minutes (Figures 1 and 2). When I was practicing yoga, I remember the instructor asking us to reach up and twist our wrists; mine crunched like Rice Krispies. The same thing happened after I iced them the first time and then continued to diminish as my wrists got well (Figures 3 and 4).

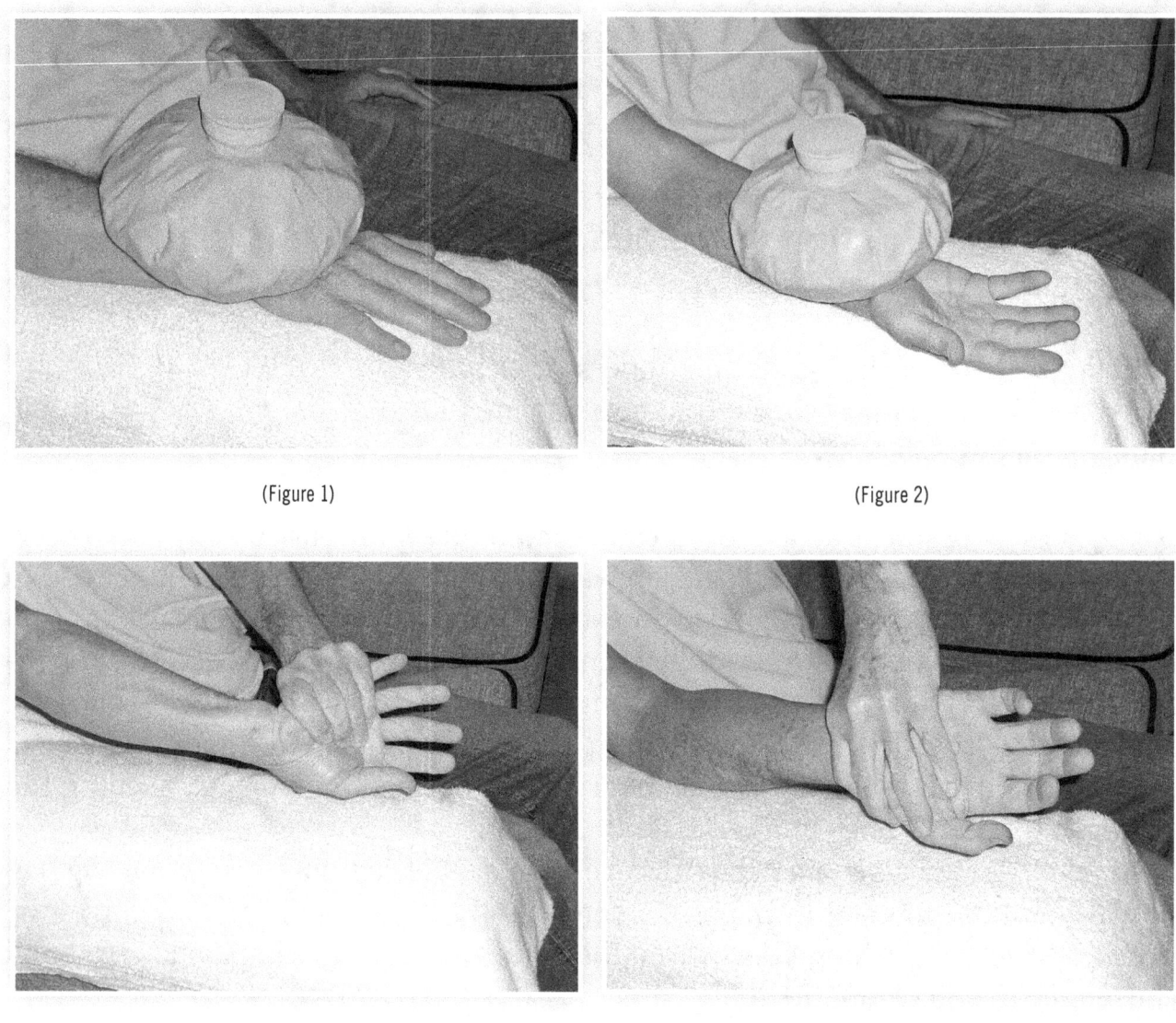

(Figure 1)

(Figure 2)

(Figure 3)

(Figure 4)

A few of the joints in my fingers began aching in my forties and I would twist them like in (Figure 5); that manipulation helped relieve the pain. After I started icing my fingers (Figure 6), it felt natural to stretch the tendons and ligaments by pulling on them (Figure 7) and by forcing the joints sideways in an opposing direction to the way they normally move (Figure 8). By icing the fingers and manipulating them in this manner, the pain in the joints of both hands has completely disappeared.

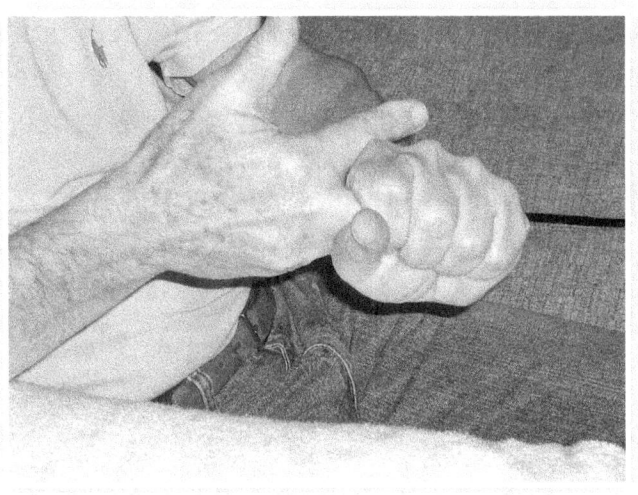

(Figure 5)

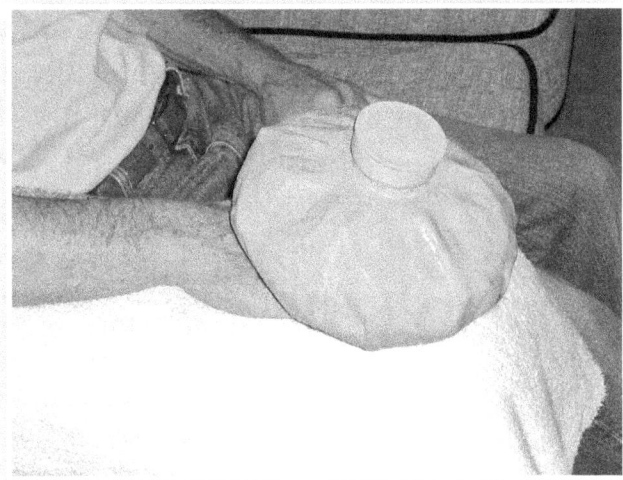

(Figure 6)

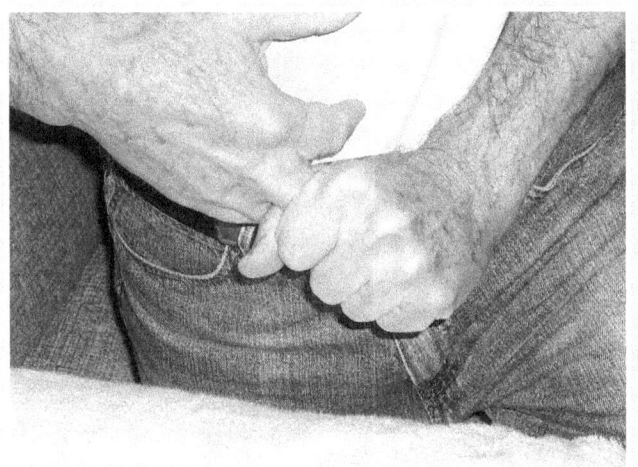

(Figure 7)

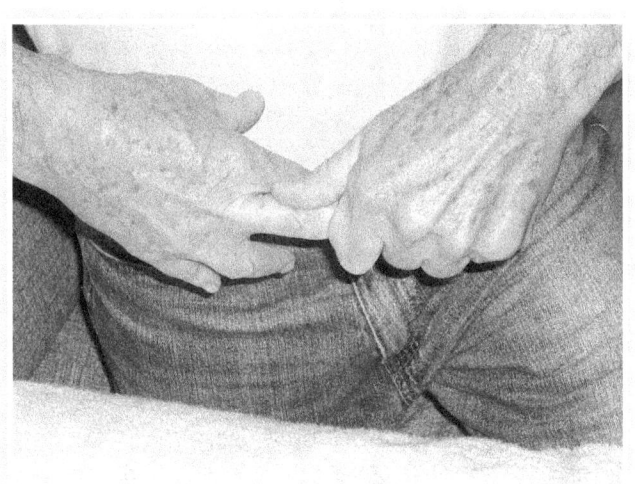

(Figure 8)

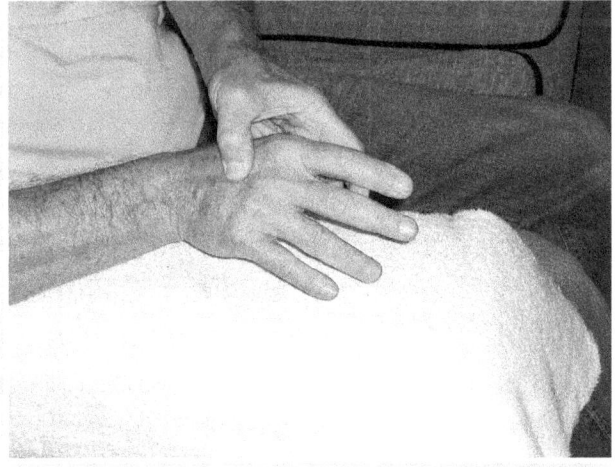

(Figure 9)

When I palpated the bones in my hands (Figure 9), they were all painful. Icing them (Figure 10) was even more painful, and it took several tries in order to ice long enough to do much good. After each icing, I manipulated the metacarpals as shown in Figure 11, and was rewarded with cracks and crunches similar to those in my fingers and wrists. What I discovered during this process is that with the ice bag flat on my hand, I was missing the first and fifth metacarpals, so I began elevating the hand laterally and medially (Figures 12 and 13) in order to cover the thumb and the pinky. I had to ice my hands several times and endured quite a bit of pain but I kept at it and eventually the pain disappeared.

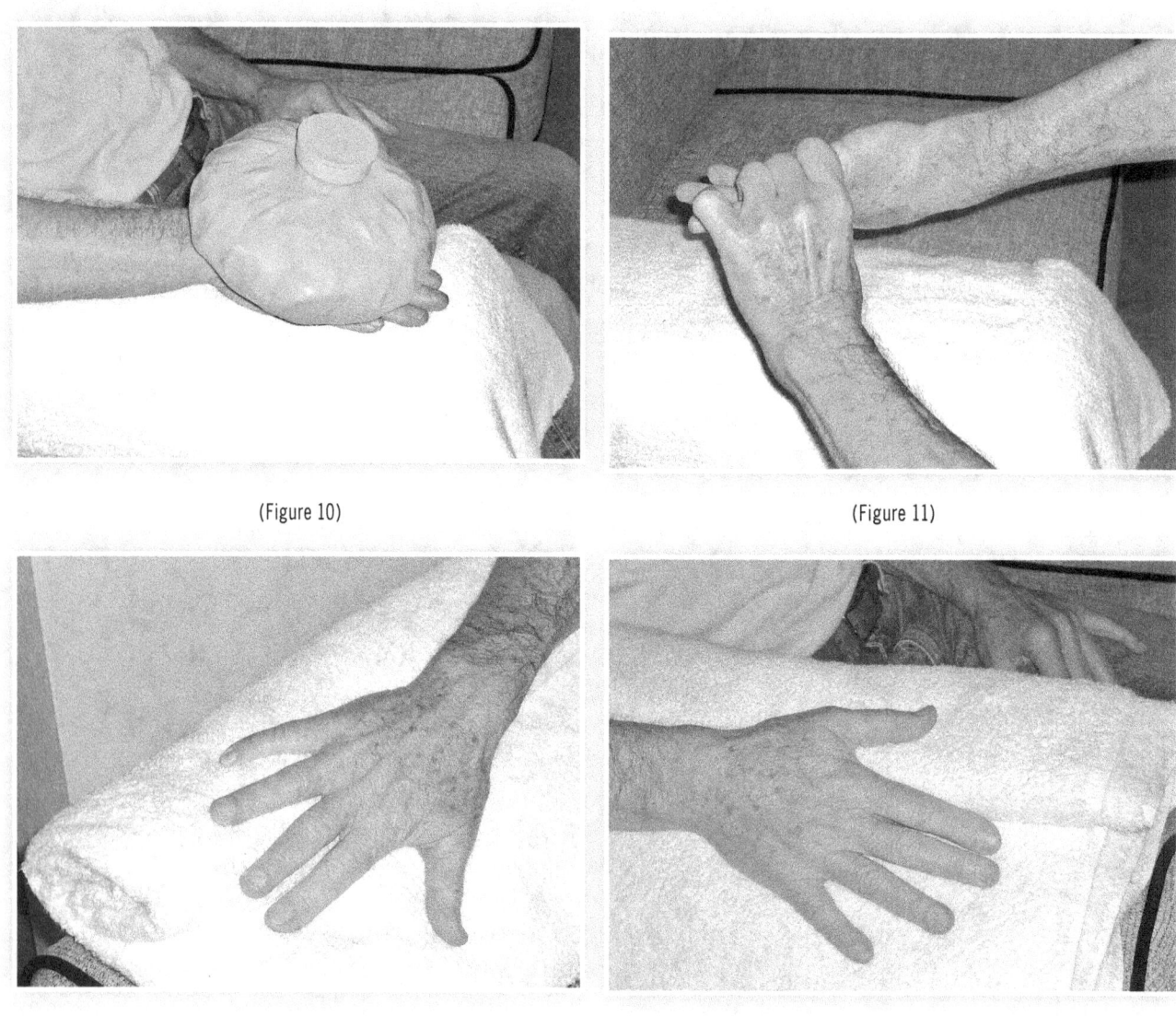

(Figure 10)

(Figure 11)

(Figure 12)

(Figure 13)

As part of the ongoing "icing one area reveals pain someplace else" saga, icing my right hand caused burning in my right forearm. I palpated here (Figures 14 and 15) and found they were sorer than a boil. Naturally, I put the ice bag on these areas (Figures 16 and 17) and, true to form, could only stand the pain for a few minutes. After the first icing, for some reason I took my right hand in my left, pushed down (Figure 18), and heard a sickening squish.

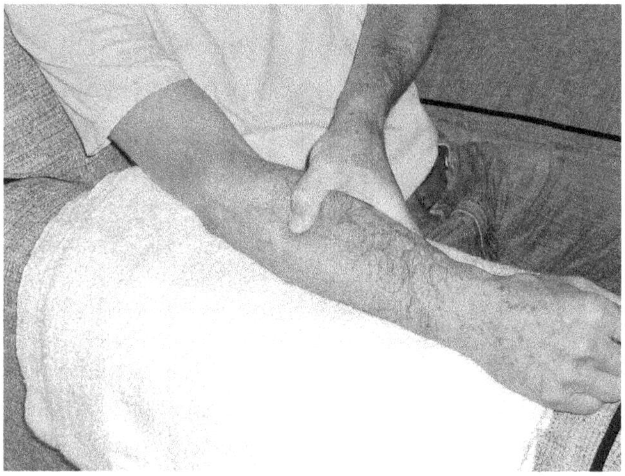

(Figure 14)

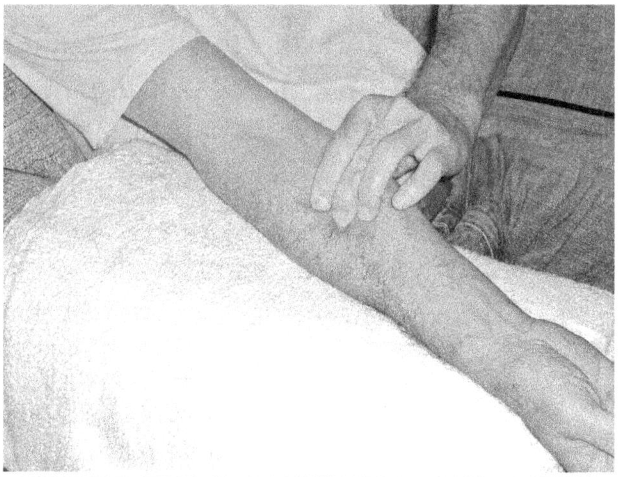

(Figure 15)

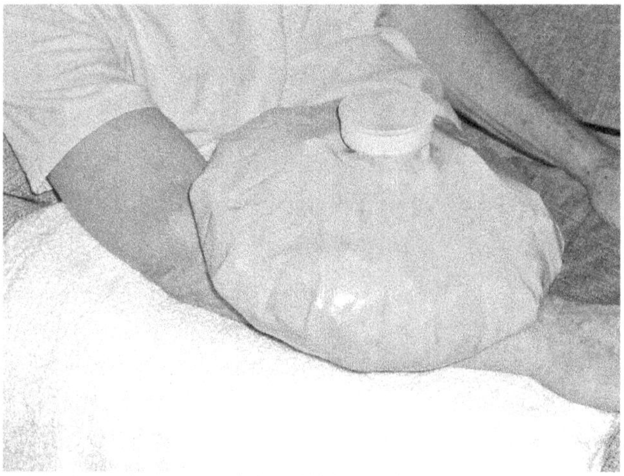

(Figure 16)

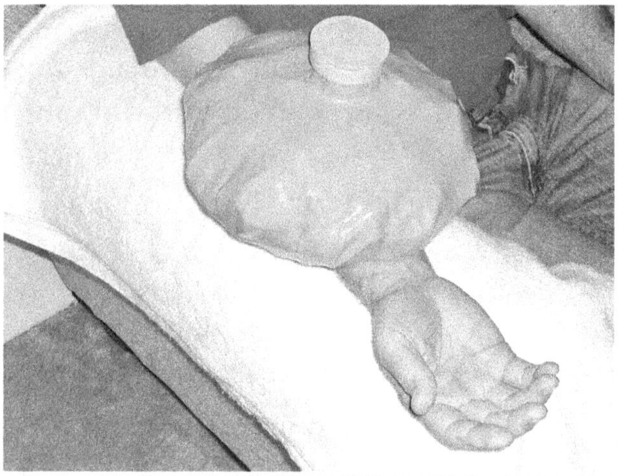

(Figure 17)

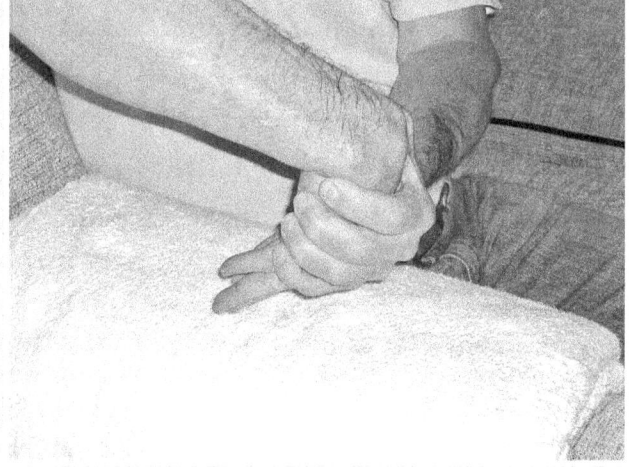

(Figure 18)

Timing is an interesting phenomenon because, shortly after, I was watching one of the international news stations and saw a report on carpal tunnel syndrome. Carpal tunnel syndrome results from compression of the median nerve in the carpal tunnel, with pain and burning or tingling in the fingers and hand, sometimes extending to the elbow. A young man was demonstrating the latest treatment. He was wearing a leather contraption that had thongs around the second, third, and fourth fingers that attached to a wide leather brace around his wrist and lower forearm; it was meant to restrict the wrist from extending, thus avoiding painful movement. After he had demonstrated the contraption, he said, "Oh, this muscle here [he pressed his thumb into his forearm at the same place as in Figure 14] is involved, too." What "the latest treatment" was keeping his wrist from doing is exactly what fixed mine (Figure 18).

Subsequent icings, and twisting and pushing down on my hand to extend the wrist as far as possible, produced more cracks, crunches, and *thunks*, and as the internal noise stopped, so did the pain. The full strength of my hands and wrists returned. Now I was able to knead dough and squeeze out a washcloth without the inside of my right wrist burning or having shooting pain in my wrists and up my arms.

Two incidents happened to my elbows, each at a different time, causing a similar result on both sides. First, I was walking into a grocery store when a dog on a leash attacked me and grabbed my right arm just below the elbow in his jaws. That required stitches, antibiotics, and a tetanus shot. The second happened when I was transplanting two citrus trees; when I stomped on the shovel to cut the last root of one of the trees, it catapulted back at me and one of the huge thorns on a branch stuck into my left elbow.

Both of these incidents caused my elbows to react much like tennis elbow, the right far more severely than the left. I'm sure they both fit the category of hidden pain because when I applied the ice bag, they both reacted with a vengeance. Before icing, I had never had a "problem" with either one of them.

The areas that I palpated for pain and then iced are shown in Figures 19, 20, 21, and 22. When icing the top of the elbow (Figure 20), you may have to ice the elbow in two positions, flexed and extended, because the muscle moves around. My pain in both elbows was so tenacious that several times during the process, I was close to seeing a doctor and asking for a cortisone shot. But I hung in there and I'm glad

I did. After icings, when I did my twisting/bending exercises, I experienced a huge realignment in my left shoulder and resolution of pain in my head and neck. Most bizarre of all, I experienced shooting pains and then the feeling of relaxation of the connective tissue in the areas of my lower abdomen and symphysis pubis (a joint made of cartilage between the right and left sides of the pubic bone, located above the genital area).

Everything is connected.

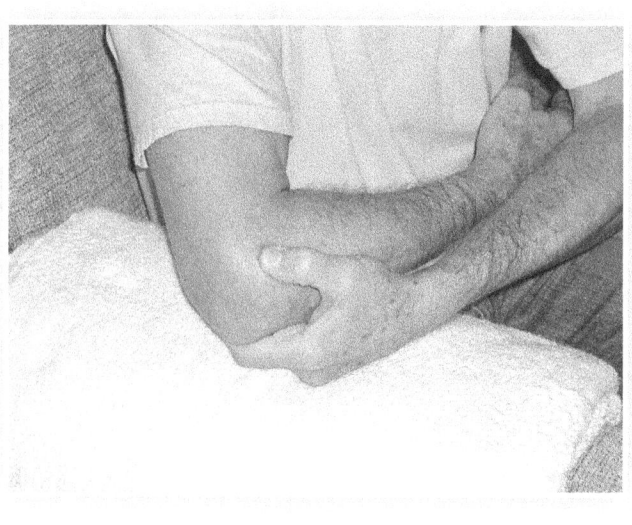

(Figure 19)

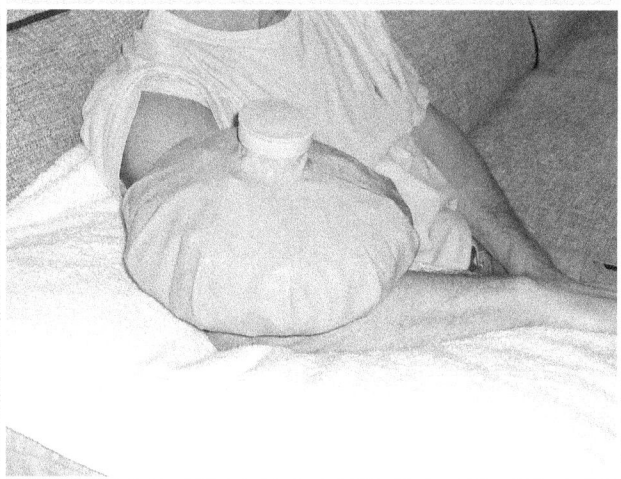

(Figure 20)

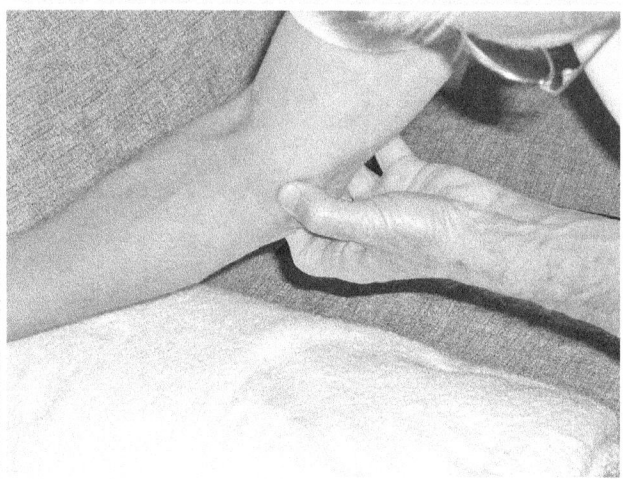

(Figure 21)

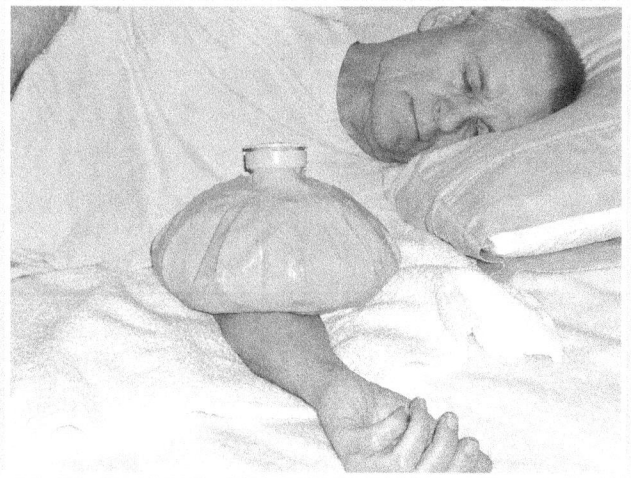

(Figure 22)

Chapter 5. Thoracic Spine
(the 12 vertebrae of the upper back);

Scapula (shoulder blade); **Ribs**
Upper back pain (irritable bowel syndrome)
Scapular pain/rib pain, including costal chondritis (inflammation of cartilage around the ribs)
Pectus excavatum (undue depression of the breastbone)

The thoracic spine is the longest of the four spinal groups and is composed of the twelve vertebrae in the upper back. These vertebrae are held firmly in place by ten "true ribs" and two short "false ribs" on either side of each vertebra. The true ribs attach to two places (facets) on each side of the vertebra and extend downward, around, and back up to become continuous with cartilage that then attaches to the sternum, or breast bone. The strength of this cage is extremely important for the protection of the vital organs of the heart and lungs.

Because the rib cage makes it very difficult for the thoracic spine to be thrown out of alignment, when it *is* thrown out, it's difficult to get it back into place. The last chiropractor I visited in the town that I lived in before I retired stated this fact to me, and he—like all the chiropractors who preceded him—was never able to realign my upper back or get any movement out of it whatsoever.

Since the thoracic cavity, the area inside the rib cage, is the longest and strongest (most supported) of the spine systems (including twelve vertebrae, twenty-four ribs, and the sternum), it took a long time and many sessions to get the soft tissue and bony structures to relax and realign. The following figures demonstrate where I found pain and how I iced it away: Figures 1 and 2; Figures 3 and 4; Figures 5 and 6; Figures 7 and 8; Figures 9 and 10; Figures 11 and 12; Figures 13 and 14; Figures 15 and 16; Figures 17 and 18; Figures 19 and 20; Figures 21 and 22; Figures 23 and 24.

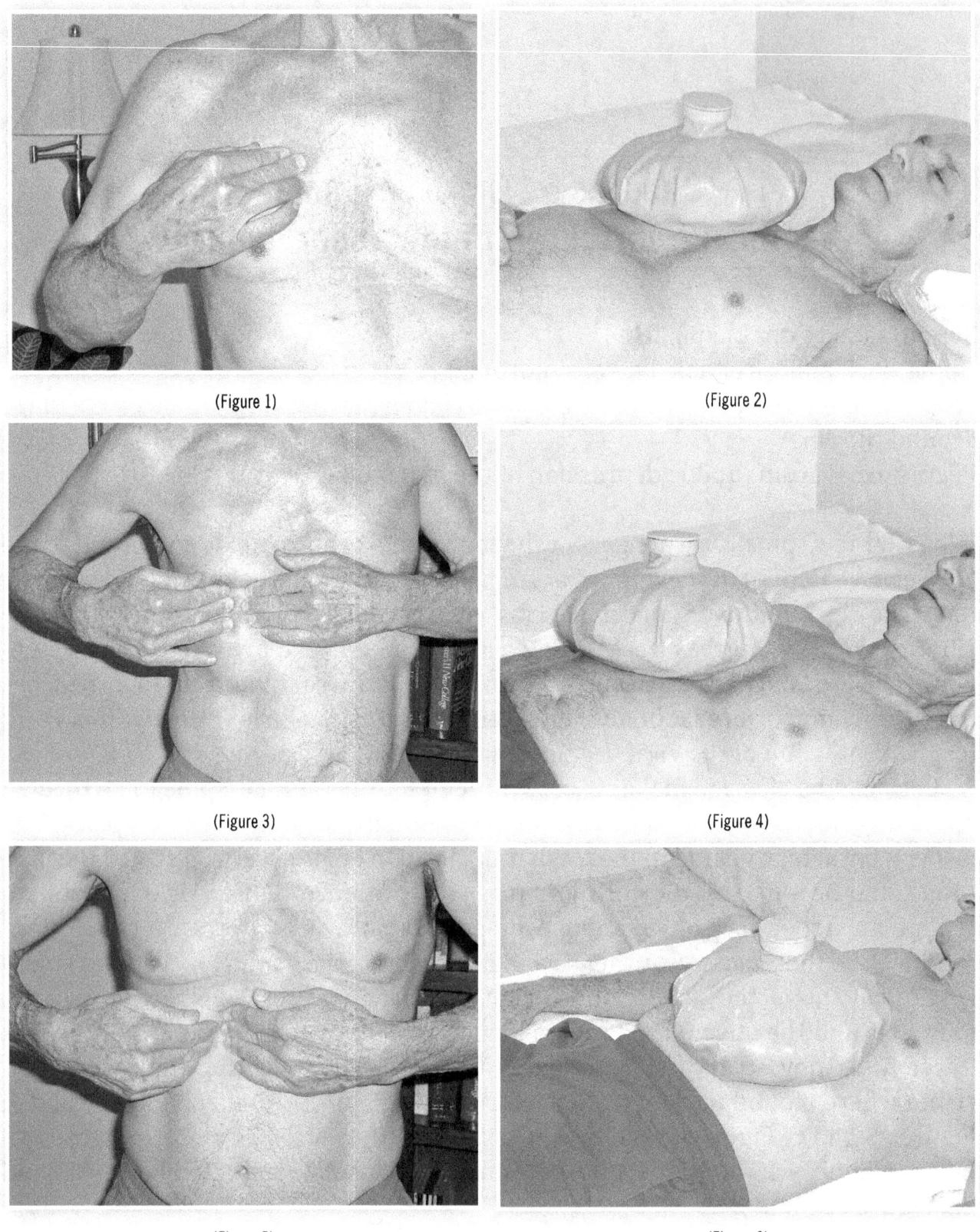

(Figure 1)

(Figure 2)

(Figure 3)

(Figure 4)

(Figure 5)

(Figure 6)

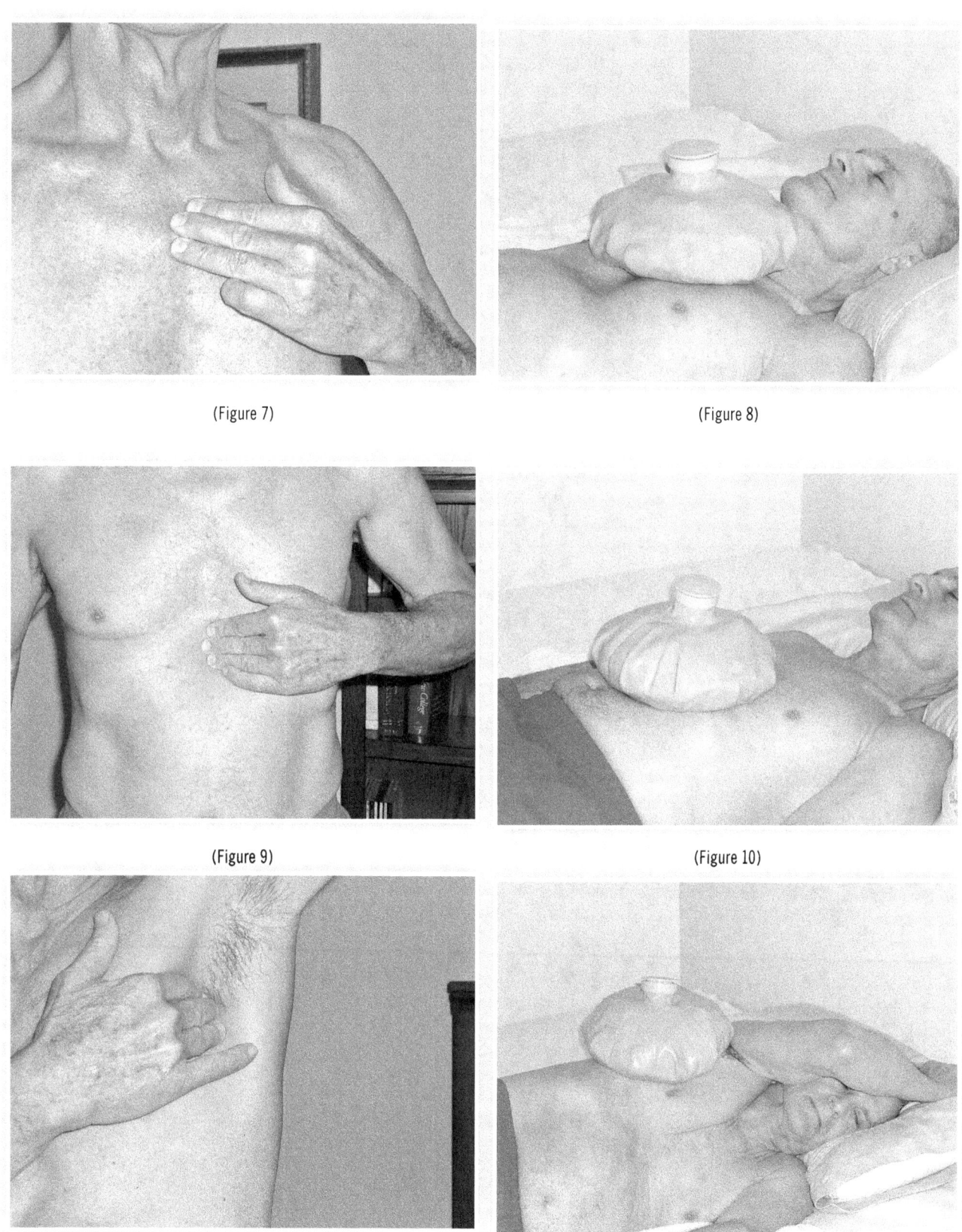

(Figure 7)

(Figure 8)

(Figure 9)

(Figure 10)

(Figure 11)

(Figure 12)

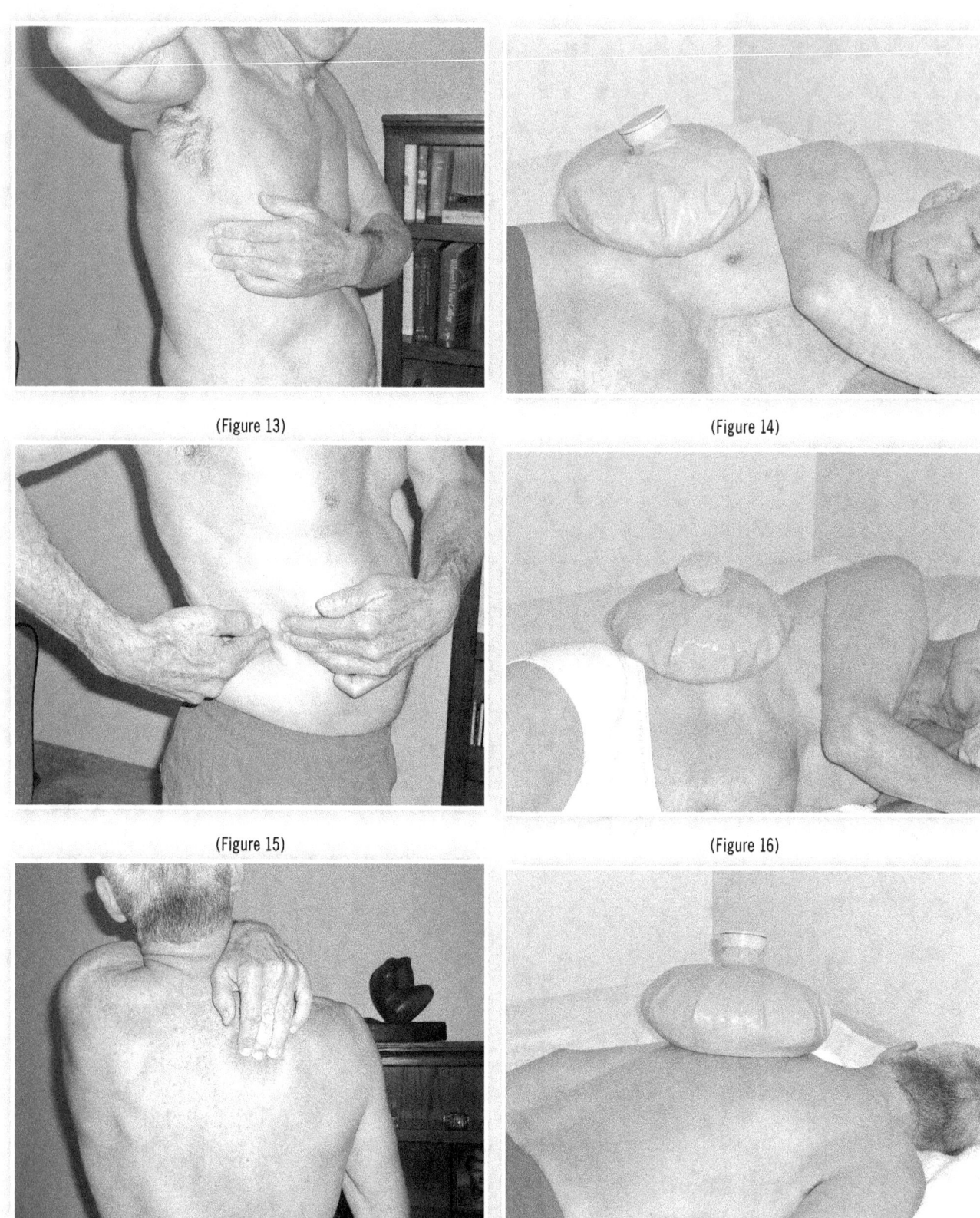

(Figure 13)

(Figure 14)

(Figure 15)

(Figure 16)

(Figure 17)

(Figure 18)

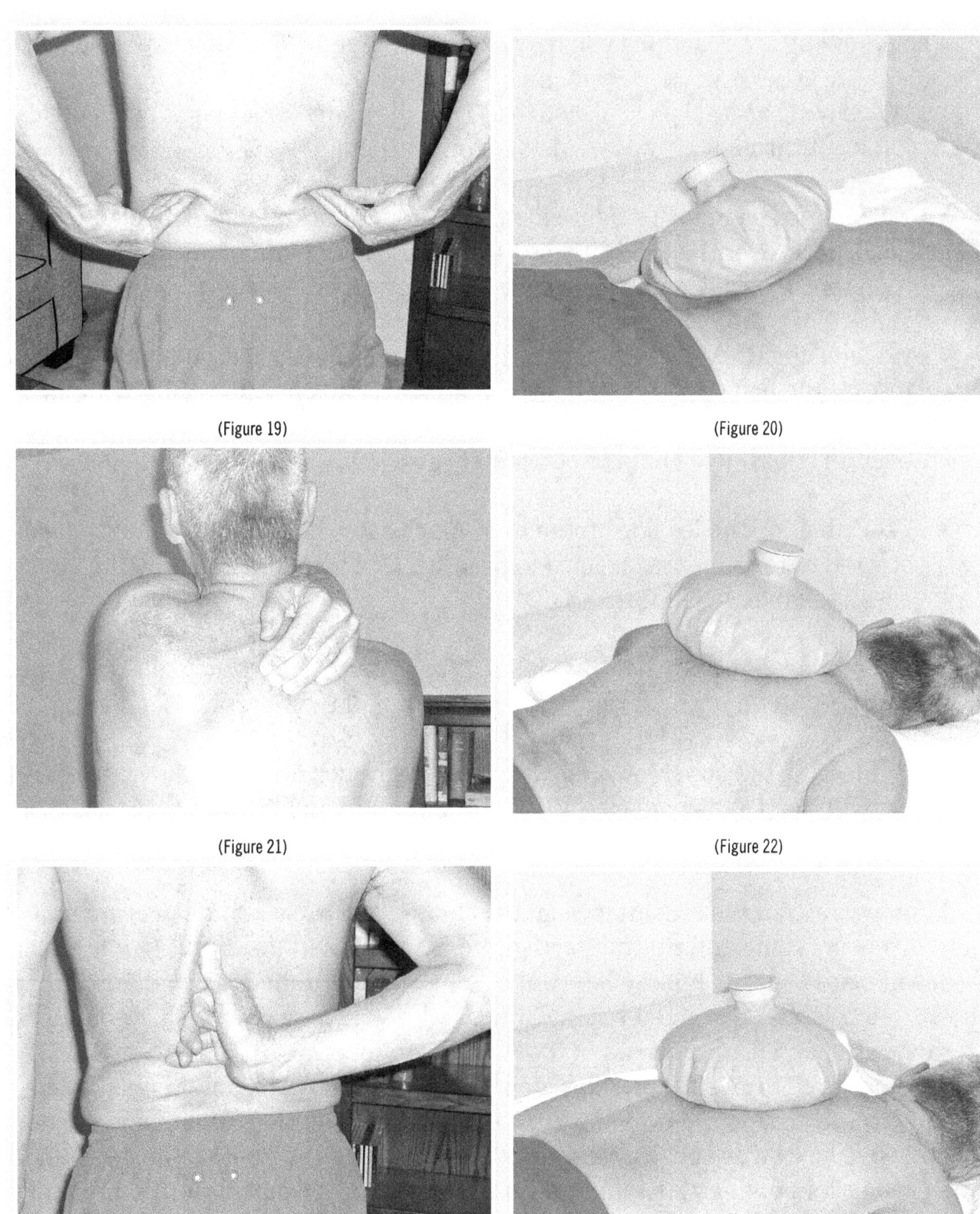

(Figure 19)

(Figure 20)

(Figure 21)

(Figure 22)

(Figure 23)

(Figure 24)

The following are reactions you may experience while icing your upper back:

- When lying on a gel pack placed in the middle of my upper back, an electrical jolt went through my body, raising it off the bed.

- With an ice bag on the right side of my upper back, a pain traveled up my back, neck, over the top of my head, and down the front of my forehead, and then the right side of my face went numb for several seconds.

- When using the gel packs in the middle of my upper back, suddenly I felt as if someone had stuck a needle in my right eye. I went into the bathroom expecting to see that a blood vessel had broken, but the eye was clear. It took over an hour for the pain to completely go away.

- While having the ice bag on the mid-upper back, in the third hour my heart started to race and it was hard to breathe, as if someone were sitting on my chest. It happened more than once.

- There was a feeling that a wide steel band was being cinched around my chest, tighter and tighter for a few seconds, and then it slowly let go.

- While icing the upper left side of my back at the scapula, I felt an intense pain around the scapula and ribs. When I did my twisting/bending exercises, the entire left side of my rib cage shifted and sounded like shaking a skeleton.

I know now that the crushing weights I carried as an assistant cameraman not only threw my spine and ribs out of alignment and caused the muscles in my back to permanently spasm, but my internal organs were "stunned" as well. Beginning in my late twenties, I started having abnormal bowel movements that got worse as the years passed, and when I had to use the toilet, I had to use it "now!" I was defecating three to five times a day, depending on what I ate, and I began having to plan my life around the proximity of a toilet. Traveling was difficult and uncomfortable because I was not always able to be near enough to a bathroom. I assumed I had what is known as irritable bowel syndrome: a chronic non-inflammatory disease characterized by abdominal pain, altered bowel habits consisting of diarrhea or constipation or both, and no detectable pathologic change. It is also called spastic or irritable colon.

The colon begins near the appendix in the right lower quadrant of the abdomen and makes a ninety-degree turn to the right upper quadrant, then makes a left turn and crosses below the diaphragm to the left upper quadrant, where it turns downward to the left lower quadrant and connects to the sigmoid colon.

During the process of icing my thoracic spine, I experienced the complete correction of my irritable bowel syndrome when the colon in each of the four quadrants, at different times, "untwisted" and relaxed. It felt like a boa constrictor uncoiling in my abdomen. The first two quadrants happened at the forty-five-minute benchmark when I was icing with the gel packs, and the other two occurred in the third and fourth hours after I began using the ice bag. After each event, my bowel movements became less frequent and more normally formed, and I had less gas. After the colon in the last quadrant realigned itself, my regularity was completely restored. After decades of my bowels controlling me, I was finally able to control them. The sudden, uncontrollable urges were gone.

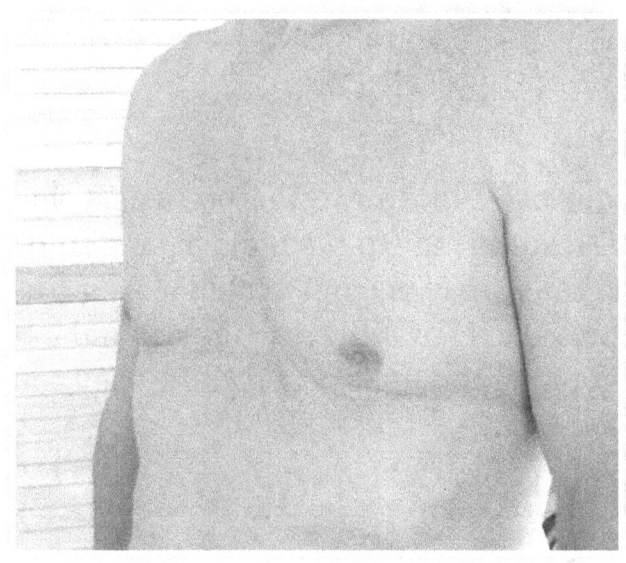

(Figure 25)

In my late fifties, during a physical, the doctor diagnosed me as having pectus excavatum (Figure 25), a condition that causes the chest to look caved in. I knew I'd had a sunken chest most of my life but, like the doctor, I didn't know what caused it. I figured it was genetic, just the way I was born. When I began icing my upper back, several things happened to change my mind. While icing, I felt my sternum drop toward the mattress, and it and the surrounding area burned. Several times I felt a huge pain start in front and work its way around to the back, and a rib would twist and move at the same time. Several times I could feel the broad muscles in my back release and relax.

After these events, the burning pain in the connective tissue between my ribs (called costal chondritis) went away. When I did my twisting/bending exercises on the side of the bed, my upper body cracked and crunched, and I could feel the vertebrae and ribs move. When I performed my roll-ups on the floor, my vertebrae moved up and down, like the keys of a piano when you run your finger across them, and I could feel bone and tissue rearrange itself. When I did my sit-ups/pull-ups with my "gut buster," (more on this in Section III Chapter 2) I felt like the

Incredible Hulk — only I didn't turn green — when a rib would turn under the skin and stick out in sharp relief then move again while it realigned itself.

As a result of ice therapy, the spasms in my upper back relaxed, my ribs re-aligned, and my sternum expanded, causing the depression between my pectoral muscles to become more normal (Figure 26).

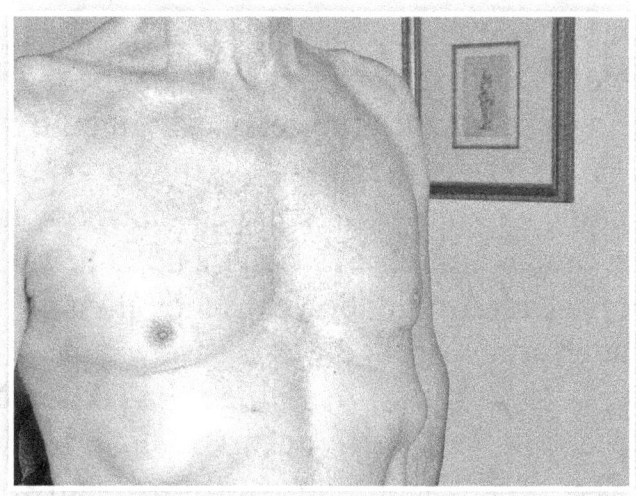

(Figure 26)

Because the scapula sticks out beyond your back, like any other body part that sticks out, it is susceptible to injury. If you're suffering from shoulder pain and/or upper back pain, consider that part of the cause may be your scapulae (Figures 27 and 28). They can be awkward to palpate and ice but if they're injured and left untreated, they can inhibit the healing of shoulder and upper back pain. They can also cause pain in your neck and head.

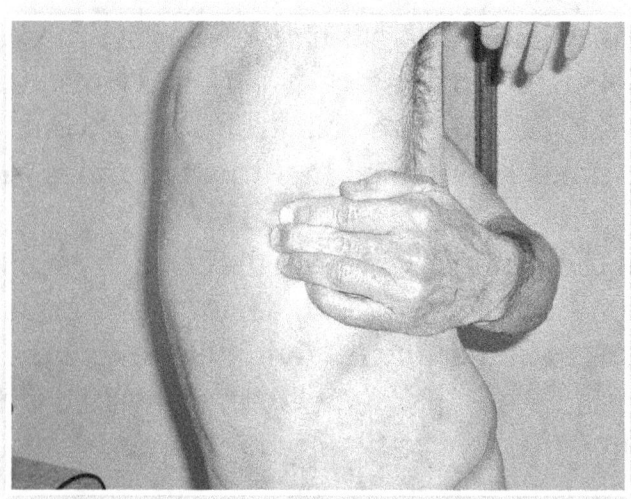

(Figure 27)

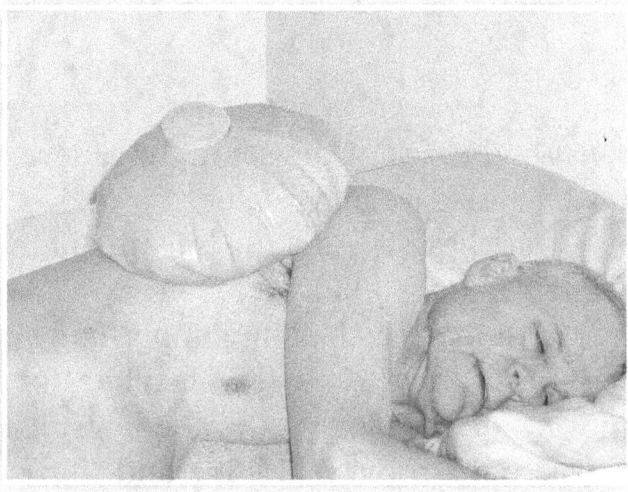

(Figure 28)

Chapter 6. Lumbar Spine (the five vertebrae of the lower back);

Sacrum/Coccyx (the triangular bone just below the lumbar vertebrae, formed usually by five fused vertebrae, and the three to five small bones below the sacrum, called the tail bone); **Pelvis**

Low back pain
Sciatica, or pain of the sciatic nerve, with spasms of pain in the buttock, back of the thigh, or in the leg or foot, following the course of the branches of the sciatic nerve
Spondylolisthesis (forward displacement of one vertebra over another, usually at the last lumbar vertebra and the first sacral vertebra)
Rectal and anal pain
Varicocele; a distention of the vein that's attached to the cord supporting a man's testicle
Nervous bladder/post-urination incontinence
Shortness of breath/vertigo

There are two areas of the spine and the surrounding tissue that are prone to injury, just because of the way we are built. These are the two junctions where lordosis (the forward curve of the neck and low back as viewed from the side) meets kyphosis (the backward curve of the upper back and buttock as viewed from the side). The last vertebra of the low back and the first vertebra of the tail bone is the junction that most people have trouble with, because it bears the weight of the entire upper body.

When I first began icing with the gel packs, I concentrated mainly on my lumbar spine (the five vertebrae of the lower back) at the junction I've described in the previous paragraph. The results were phenomenal. I couldn't believe the movement that I could feel and hear. During all the years of treatment, I heard a pop, crack or crunch in my low back (on the right side only) or in my neck when they were adjusted by a chiropractor; year after year, they were the exact same sounds in exactly the same places. I never got better; as a matter of fact, I got worse. After the "adjustment", the muscle spasms would eventually pull the joints out of alignment again, and I'd be back to square one.

Within weeks of beginning ice therapy, I could lean over the bathroom sink to brush my teeth without holding myself up with one hand, and lean over to pick things up without my back going out. However, as time passed, the improvement stalled and it became apparent that my low back pain wasn't coming from the vertebrae in the lumbar spine alone. I probed and pressed the surrounding muscles and found that almost every area was painful. In my case, the majority of my pain was caused by inflamed and swollen soft tissue (connective tissue), and until that was healed, the pain didn't go away.

The following are areas of injured connective tissue and their attachments that caused my low back pain and other ailments associated with it. I suggest that you try the following based on what I did to peel away the layers of pain.

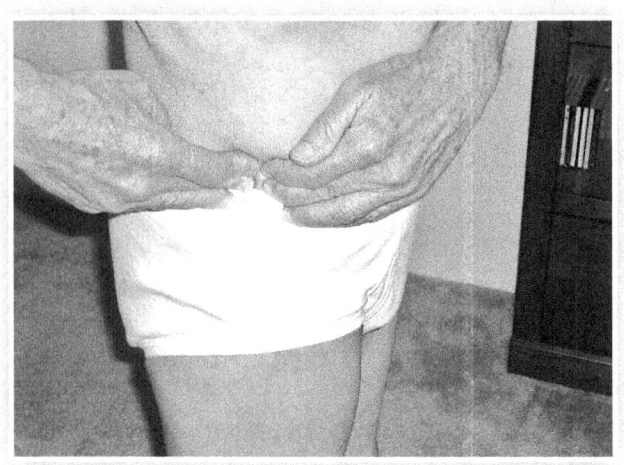

(Figure 1)

Palpate the top of the hip bone (Figure 1), and the lower margin of the ribs (Figure 2); if the area is painful, center the ice bag here (Figure 3). Press hard in the muscular area below the iliac crest (Figure 4) and center the ice bag here (Figure 5). Moving around to the lower back area, press hard on this muscle at the level of the iliac crest (Figure 6) and also at the lower margin of the ribs (Figure 7). Place the bag here (Figure 8). Next, palpate the gluteus maximus muscle (Figure 9) and also the sits bone area (Figure 10), and place the bag here (Figures 11 and 12). Then zero in on the spine itself by pressing hard on the outside edges of the lumbar spine (Figure 13) and the area near the junction of your low back and sacrum (Figure 14). Center the ice bag here (Figure 15). Check the sacrum area (Figure 16) and the coccyx (Figure 17), and center the ice bag as shown in Figure 18.

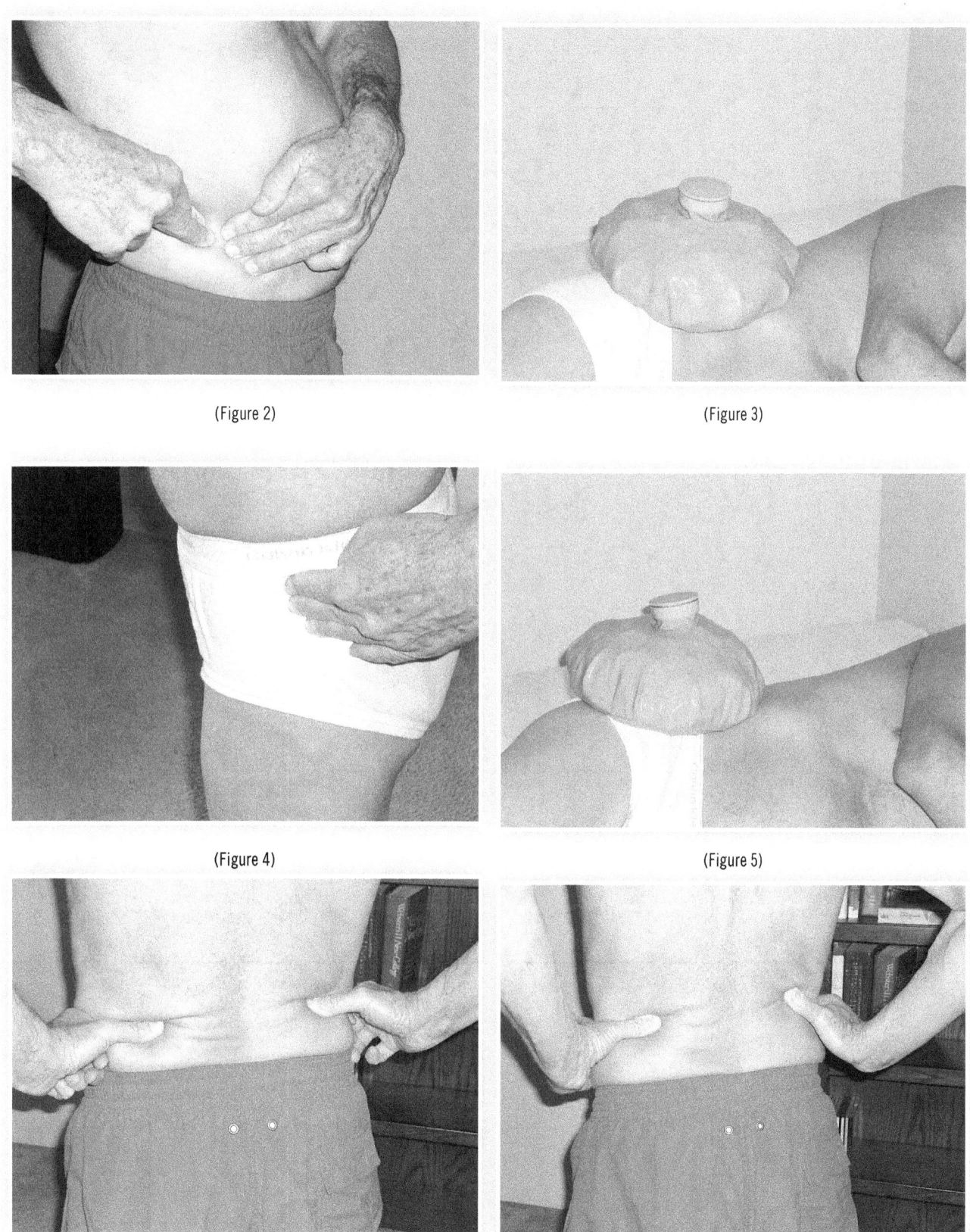

(Figure 2)

(Figure 3)

(Figure 4)

(Figure 5)

(Figure 6)

(Figure 7)

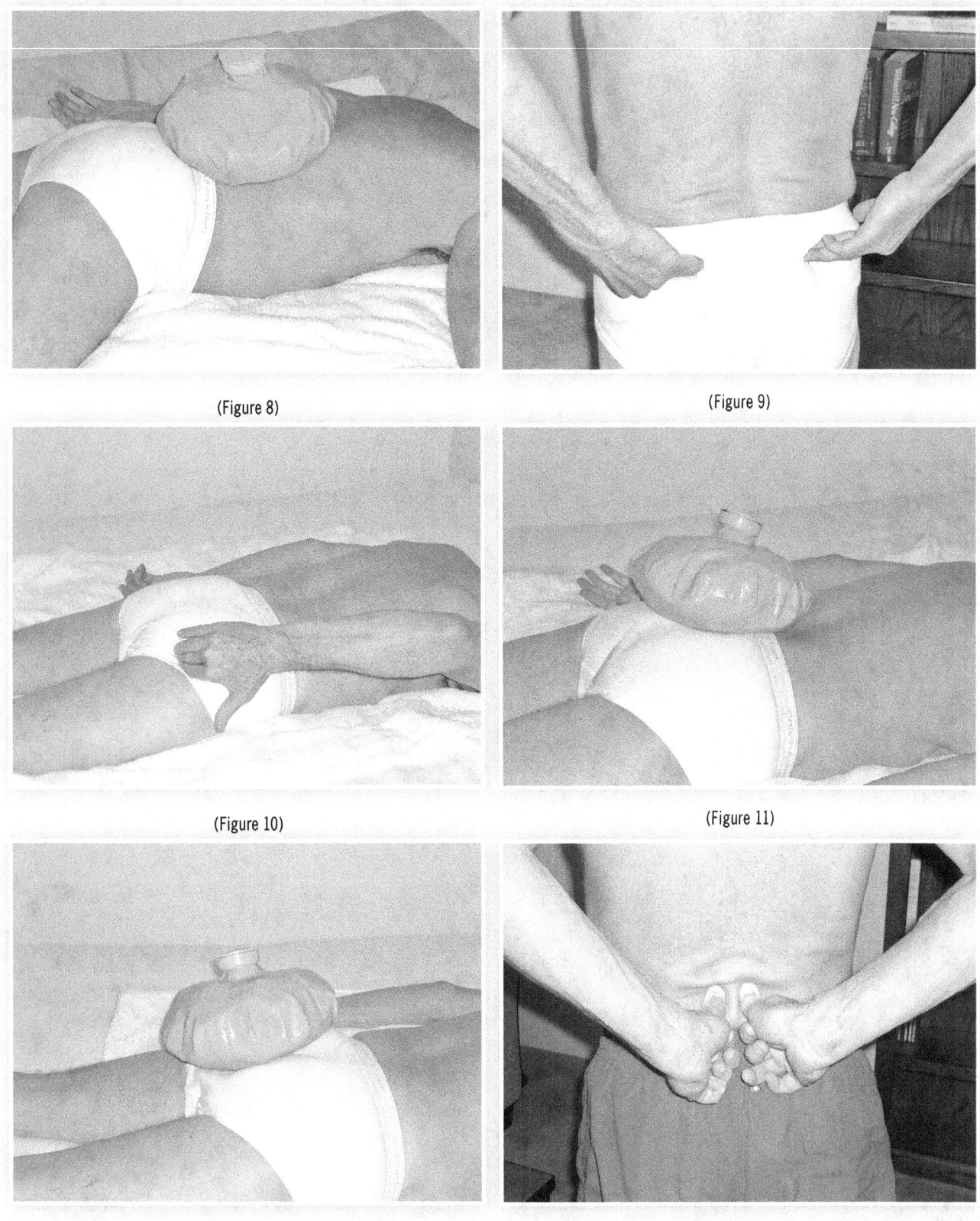

(Figure 8)

(Figure 9)

(Figure 10)

(Figure 11)

(Figure 12)

(Figure 13)

it subsided. That was one of those times when I was sure the pain was gone for good — and it was.

Many years ago, an X-ray had revealed that I had a spondylolisthesis, meaning a forward displacement of one vertebra over another, leaving the last lumbar vertebra hanging over the first sacral vertebra (Figure 19). After I had been icing for a few months, I was warming up for a yoga class and was doing a pose called the "plough" (Figure 20). I had held the pose for a few minutes and when I brought my legs back up over my head, I heard and felt my pelvis shift forward with a huge *thunk*. I was shocked into immobility and all I could do was lie there with my knees up, but I knew what had happened. The next day at work, I asked one of the X-ray techs to take a spot film of the two vertebrae involved (L-5 and S-1), and sure enough, they were in perfect alignment (Figure 21).

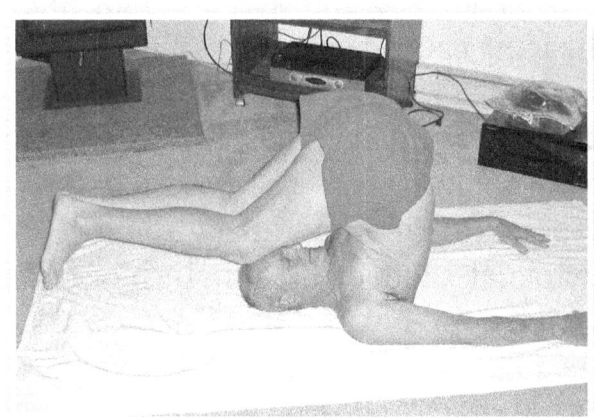

(Figure 20)

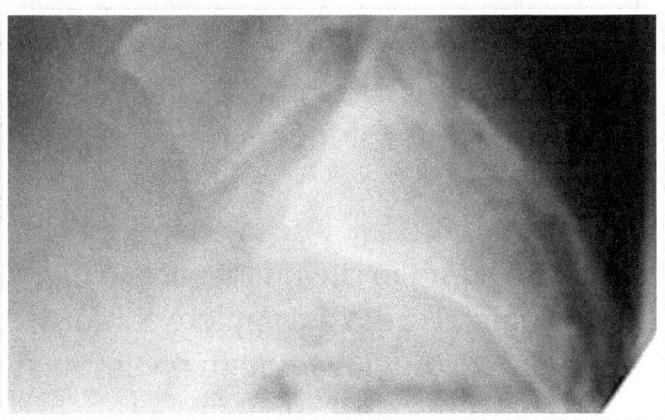

(Figure 21)

As the days went by, I expected the level of pain in my back to decrease, but there was no change at all. This more or less confirmed my suspicion that most of my back pain was coming from swollen, inflamed muscles and irritated nerves rather than vertebrae being out of alignment. (Please refer to the exercise information in Section III, Chapter 2 before attempting the "plough.")

In my late twenties, after I'd been carrying cameras for several years, I started experiencing rectal and anal pain. Having a bowel movement was excruciating and if the stool was hard, it felt as if someone were putting a knife in me. The same pain was triggered after a climax as well. Sometimes I would wake up with the pain in the middle of the night, usually with an erection, and the only thing that would stop the pain was to get up, sit on the toilet, and relax until it subsided.

I mentioned the pain to two different GPs, who both ignored it by changing the subject. These times were frustrating for me, but if a doctor won't talk about it, what can you do? In my early fifties, I had a sigmoidoscopy and when the doctor came back in the room, he said, "Well, your colon's in spasm but everything else is fine." Then he turned and walked out the door. These doctors were all older, "straight" white men, and although they didn't say so, and because of their refusal to discuss it, I was pretty sure they had preconceived ideas as to what was causing my rectal and anal pain. Because they were socially uncomfortable, which had absolutely nothing to do with me, they chose to ignore my pain.

Early on in my research, when I was lying on gel packs and experimenting with icing my glutes to see if I could get some relief with my back pain, the therapy caused shooting pains in my rectum (the lower portion of the large intestine, just above the anus) and anus. After I switched to the ice bag and was icing for hours instead of minutes, the "healing pain" that the icing caused encouraged the spasms in my glute muscles to gradually release and relax, and after years of enduring excruciating pain in my rectum and anus, almost on a daily basis, the pain disappeared completely.

In my late fifties, I began experiencing frequent urination during the day and having to get up in the middle of the night to urinate. I made an appointment with my urologist. He did a complete examination first and noticed that I had a varicocele in my scrotum leading to my left testicle. (That's a distention of the vein that runs next to the cord supporting the testicle). My family doctor had discovered it when I was fifteen and treated it with a series of four hormone shots given to me over a several-week period.

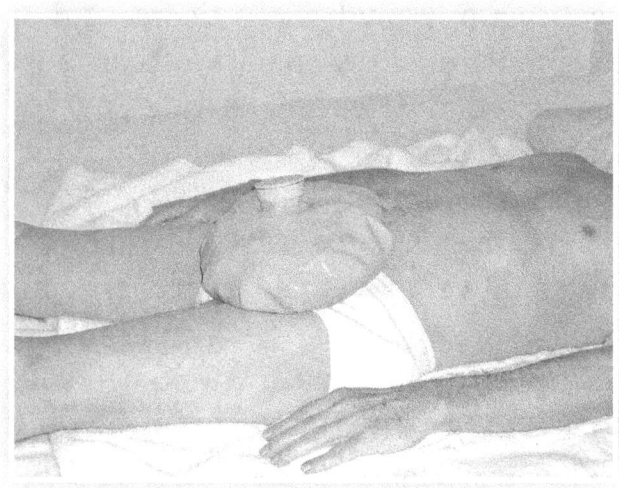

(Figure 22)

My urologist asked me why he hadn't operated on it at the time, and told me it was too late to do anything about it now and, that as a result of the varicocele, the left testicle was smaller than the right. As part of the physical examination, he took blood samples that showed a very low testosterone level of one hundred and sixty-six (normal is from one hundred and ninety to eight hundred), which I'd probably had all my life due to the interruption of blood flow to the left testicle.

When I was icing my lower back on the left side, in the third and fourth hours I began to feel pain in the left inguinal area (midway between the center of the body and the groin). The pain felt as if I were having an attack of appendicitis, only on the wrong side. After several times, I began to think that maybe an injury in front was causing pain in back so I put the ice bag on the left inguinal area (Figure 22), and exactly forty-five minutes later (eerie), an intense pain started in my left testicle and radiated up into the lower left quadrant of my abdomen. My left testicle felt like it was being sucked up into my abdomen, but after a few moments, the pain slowly receded and the testicle relaxed. I got up immediately and checked for the varicocele. It was gone.

Two years later, when I was back in the States, I made an appointment with my urologist, and told him what had happened. He examined me and confirmed that the varicocele was gone. He said, "Where did you put the ice bag? How long did you leave it there?" I explained what I had done and he said that from then on, he was going to prescribe icing the inguinal area for any of his patients who had varicoceles. After all, he said, "It can't hurt you." He ordered another blood test for PSA and testosterone levels, and my testosterone level had gone from one hundred sixty-six to four hundred and forty.

As it turned out, my varicocele was nothing more than a strain that could have been corrected fifty years ago if only the doctor had known about the non-invasive healing power of ice therapy. And the fact that after fifty years my testosterone level went back to normal, blew me away. My urologist was wrong when he said it was too late to operate. After the blood flow was restored to my left testicle (no matter how it was done), the testicle began to function normally again.

My frequent urination was caused by an enlarged prostate, so my urologist performed laser surgery that eliminated some of the prostate tissue surrounding the urethra and allowed my bladder to completely empty. That solved my frequent urination problem but didn't do a thing for the sudden urges. Like my irritable bowels, my irritable bladder let me know that when it wanted to go, it wanted to go "now."

When I began icing my low back at the juncture of my lower spine and the sacral area (Figure 15) for long periods of time, in the third hour my bladder would ache and throb. While icing my low back, I would perform the maneuver described in the exercise section (Figure 10); that, along with the ice therapy, corrected the spasm in my bladder and my urination returned to normal. This is how you do it:

Take in a deep breath and hold it for a few seconds. Slowly let it out while you push down on your forearms, lifting your upper body off the bed and, at the same time, pulling your navel in toward your spine, forcing all the air out of your lungs. Keep pulling your navel in and forcing the air out then bear down on your lower abdomen. Take short breaths in order to keep yourself in that position and keep the pressure on the lower abdomen, until the pain in your bladder and inguinal area subsides, and you feel it relax.

Performing this maneuver during and after icing my lumbar spine, thoracic spine, and surrounding areas caused the correction of my irritable bladder. During one of the few times that I've iced for six hours, the left sciatic nerve (finally) let go in the fourth hour; in the fifth hour, pain in the inguinal and bladder areas disappeared; and in the sixth hour, my heart began to pound and I could feel my descending aorta in the thoracic and abdominal cavities expand and contract with my heartbeats. It lasted for only six or eight beats, but I was seriously concerned that if there was a weakness in the aorta it might burst.

After it was over, I felt like I was in limbo and thought, *Am I dead?* Slowly, my senses returned to normal. Since that experience, I have more stamina and less shortness of breath after exerting myself and, I no longer become dizzy when I lean over and the level of my head is lower than the level of my heart.

Another disturbing condition in my urinary tract began a few years ago. After I finished urinating and had put myself back in my pants, a flood of urine would come out and wet my underwear. This was annoying when I was alone and embarrassing if I was in public. I chalked it up to old age and tried to be as careful as possible after using the bathroom. One afternoon, I was icing pain in my scalp on the back of my head (Figure 4 in head section); in the third hour, I felt a mild tingling sensation in my rectum and the sphincter muscles of my anus (odd). In the fourth hour, between my legs I felt the root of my penis move like a snake uncoiling. After that, the post-urination was gone and it's never returned.

Chapter 7. Hips/Thighs/Knees

Pain/malalignments

Fibromyalgia (pain in the fibrous tissues — for example, muscle and connective tissue)

I started working as an assistant cameraman in the late 1960s, before circuit boards and Panavision, and I was carrying the old heavyweight cameras. Several times, I felt my spine shift and, once, my right hip. Many of the injuries I've talked about so far were directly related to the crushing weights I carried during the twelve years I worked as an assistant.

After I had only been icing a few weeks, I was doing my hip-stretching exercises (Figure 7 in exercises) and when I put my right foot on my left knee and pushed down, my hip joint made a squishy, sucking sound; I felt a sharp pain and my knee dropped toward the floor. At that point, I was mainly targeting my low back, but because of the reduction of the inflammation and swelling in my back, a misaligned hip I had been walking on for decades easily fell into place when I performed that simple exercise.

When I was taking X-rays for a group of orthopedic doctors, I told one of them about the pain running down the sides of both of my thighs. This doctor diagnosed it as fibromyalgia (pain in the fibrous tissues, like muscle and connective tissue) and, for several weeks, injected them with lidocaine, a local pain killer. He explained that it wasn't the lidocaine but the needle that (in theory) disrupted the electrical impulses the nerves were sending to my brain that were triggering the pain. The treatment was unsuccessful.

When I began to ice, the outside of my left thigh was the very first place I put a gel pack and, within minutes, an electrical shock shot down my leg to my foot and up my side to my shoulder. I knew in that moment that I had been exposed to an incredibly powerful healing agent. Find the pain in your thighs by palpating the attachment of the muscle to the greater trocanter of the femur, a broad, flat place at the upper end of the femur, to which several muscles are attached (Figure 1) and below that (Figure 2), and ice them (Figures 3 and 4). I have also detected pain in the front, inside, and back of my thighs, and have iced them as shown in Figures 5, 6, and 7. Spasms in the glutes and the muscles in your haunches (hips and buttocks) can also be a source of the pain down your thighs.

In my forties, I began noticing pain in my left knee and when I got down on my hands and knees to clean up a spill or to garden, it felt like there was a knot under the skin; when I put my weight on it, it burned (Figure 8). When I began practicing yoga, the knee was so painful that there were certain poses I couldn't perform.

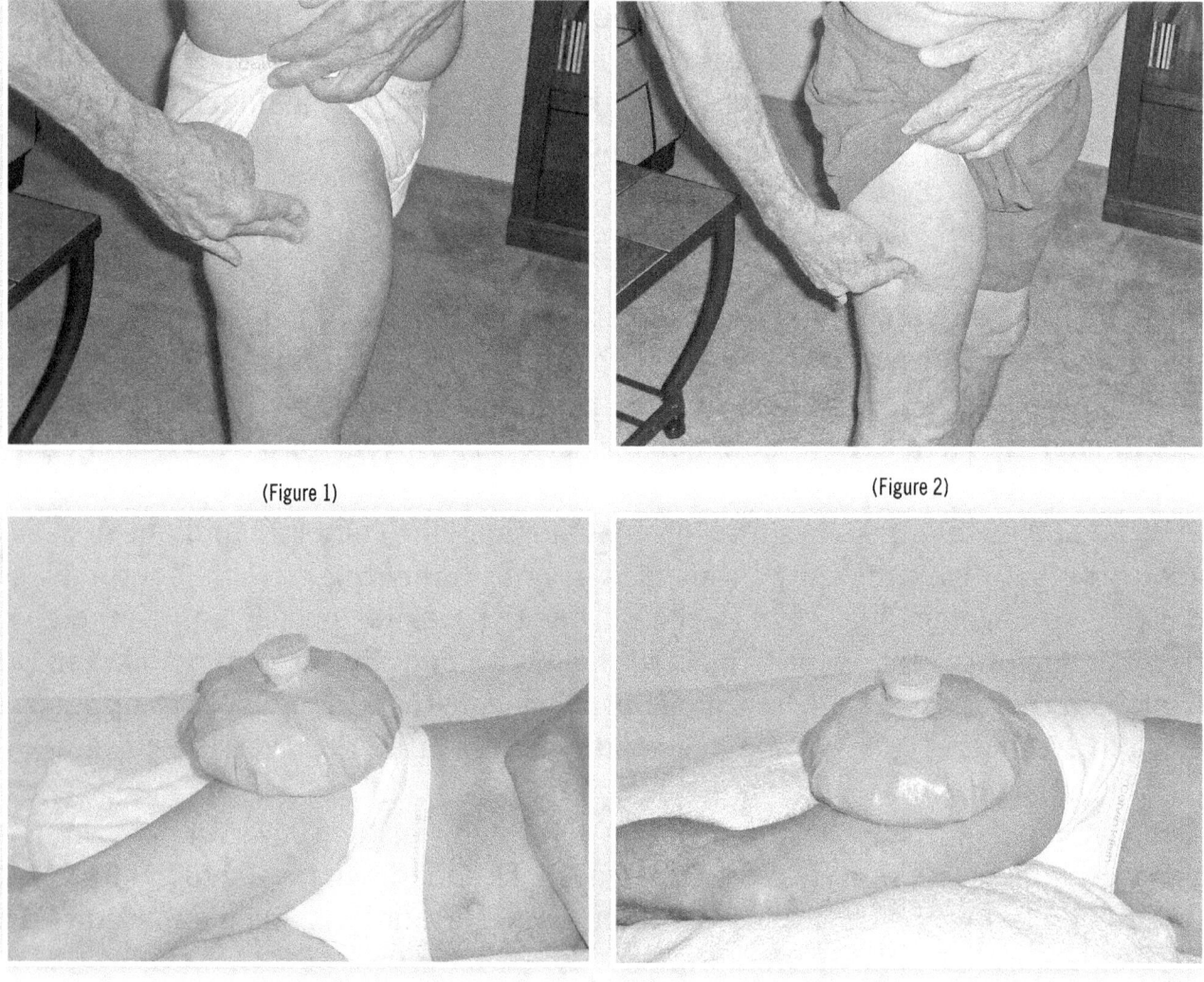

(Figure 1)

(Figure 2)

(Figure 3)

(Figure 4)

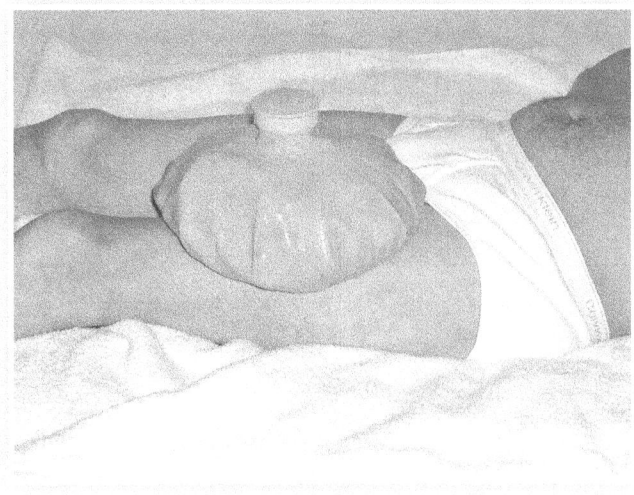

(Figure 5)

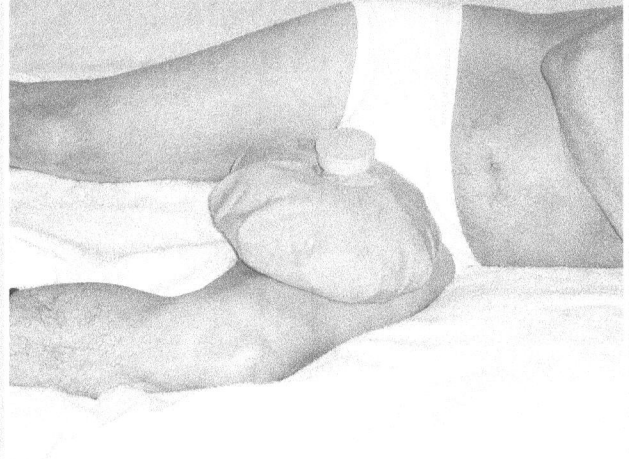

(Figure 6)

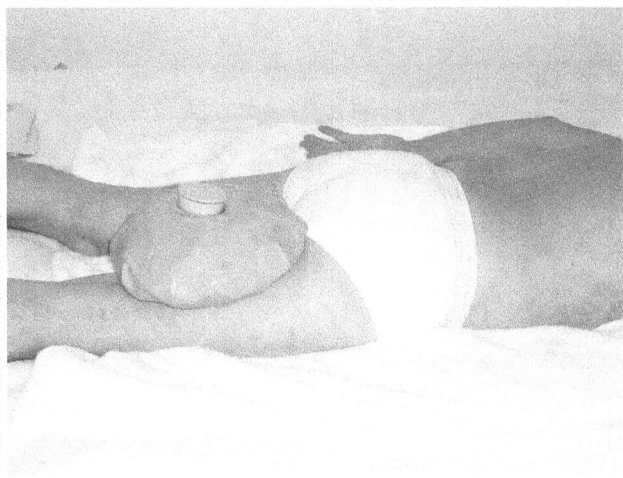

(Figure 7)

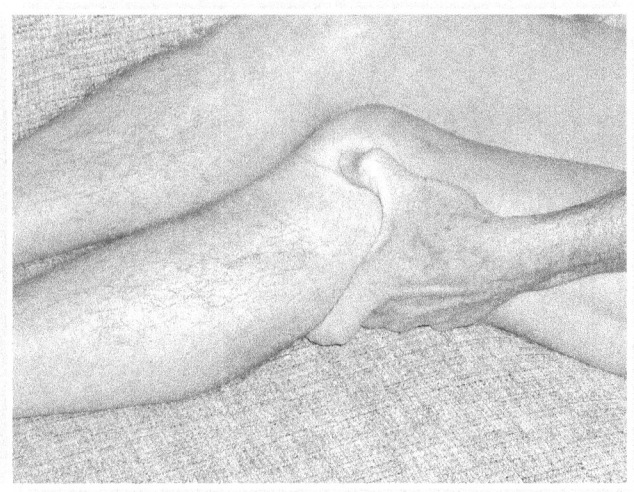

(Figure 8)

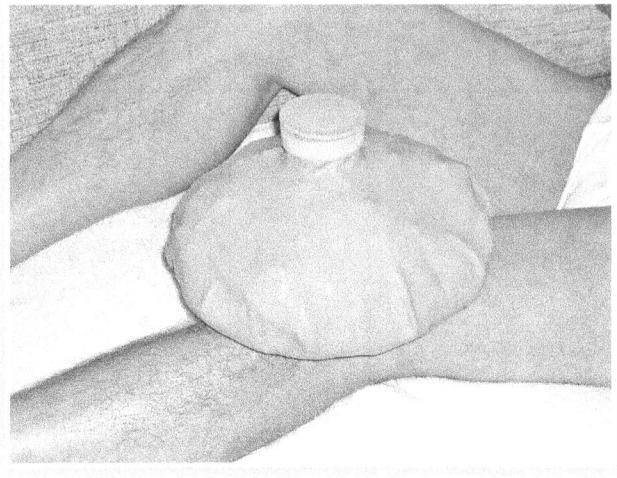

(Figure 9)

I was using the gel packs at the time so I laid one on my left knee (Figure 9) and, within twenty minutes, it was so painful that I had to take it off. What's more, after I took it off, the pain got worse so I ended up getting in the shower and running warm water over the knee to get the pain to stop. After several icings, I was able to keep the ice in place for over an hour and eventually the knot and the pain associated with it went away.

Later, I discovered painful areas around my right knee and I iced them away by placing the ice bag on the areas shown in Figures 10, 11, 12, and 13.

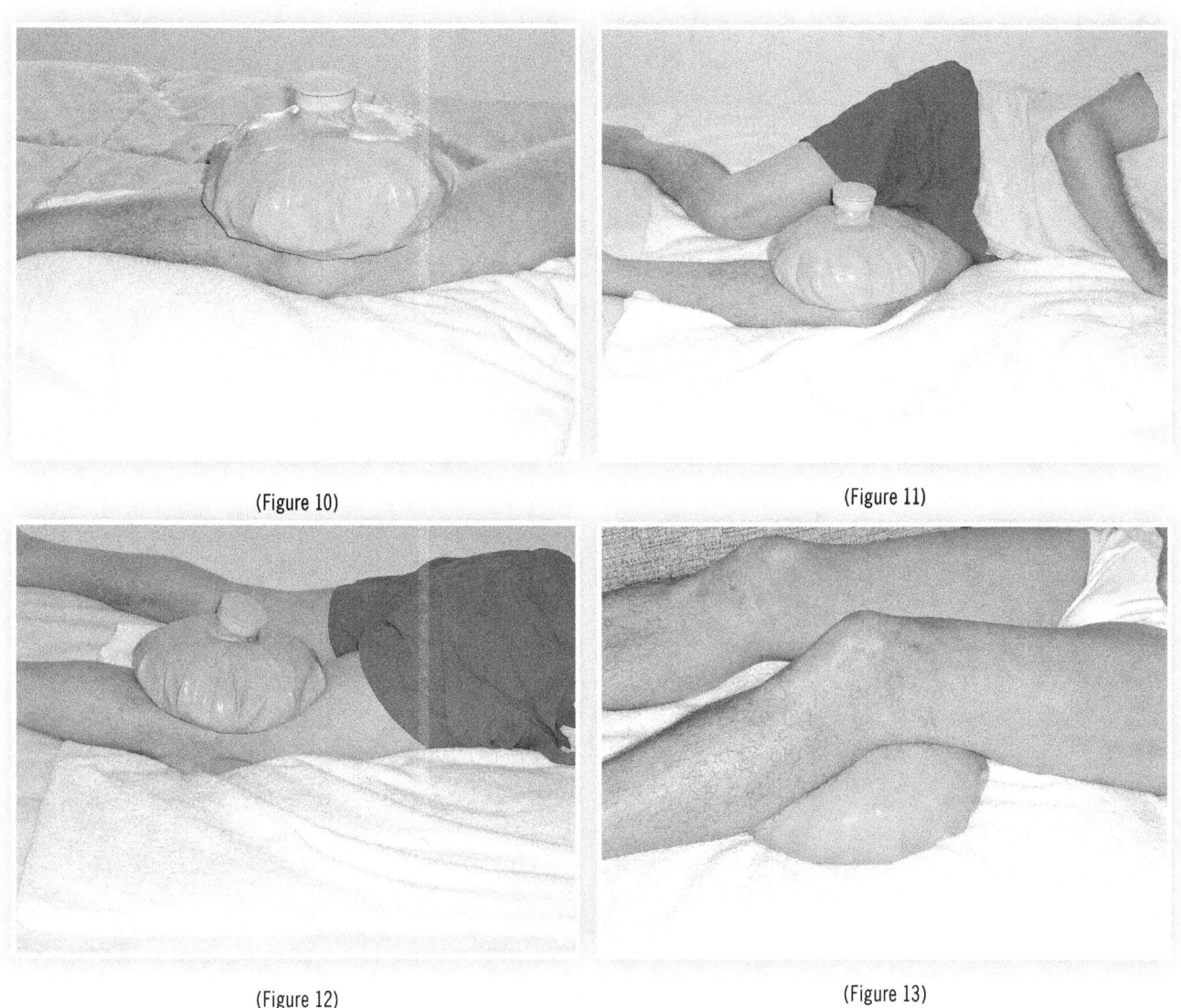

(Figure 10)

(Figure 11)

(Figure 12)

(Figure 13)

Chapter 8. Lower Legs; Ankles/Feet and Toes

Pain/spasms
Neuropathy (numbness), unrelated to diabetes

The most confounding injury I've had, one I've lived with for nearly thirty years, is numbness in the bottom of my left foot. I consulted two different neurologists several years apart; they performed identical examinations. Both doctors ran a spur-like device up the bottom and top of my foot and in front and behind my lower leg. Both their diagnoses were the same: the numbness was either caused by a pinched nerve in my lumbar spine, or by alcohol abuse.

Years later, when I was working for an orthopedic group, one of the doctors performed an electrical conduction study and said that the numbness in my foot was caused by a pinched nerve in my back; he prescribed epidural injections. The pain-management doctor scheduled six injections but, after the fourth one, told me he didn't want to continue, because he didn't think the numbness was coming from my back. At that point, I had no choice but to give up and live with it.

When I first began using ice therapy, I was using the gel packs and working mainly on my low back. Every place I iced caused my foot to react with burning, cramping, or a feeling of pressure. Since I had been told that the numbness was coming from my back, and since my foot reacted when I iced my back, I couldn't give up that idea. But after I began using the ice bag and my foot reacted to icing my head, for example, I began to get suspicious and finally started to entertain the notion that maybe the numbness was actually coming from the foot itself.

One day I reached down and started pressing and squeezing the metatarsals in my left foot; I was startled to discover that they were so painful I could barely touch them. Then I squeezed my Achilles tendon (the tendon that attaches to the back of the heel and runs up the back of the lower leg) and other areas of the foot and lower leg — and almost every area was painful.

Thinking back, I decided the auto accident I'd had in my mid-thirties, when the engine was sitting in the driver's seat, must have crushed my foot and lower leg, and done massive nerve damage. I also remembered waking up in intensive care and feeling excruciating pain in my left foot; looking down at it, I'd seen that it had been tightly wrapped in an ace bandage. I had asked to have it loosened but no one came. After being ignored for almost twenty-four hours, I stopped a man walking through my room and said, "Sir, the bandage on my left foot is killing me. I've asked that someone loosen it but no one will do it. If someone doesn't come in here in the next five minutes, I'm going to rip these tubes out and walk out of here...and if you don't think I can do it, just watch me." That did the trick.

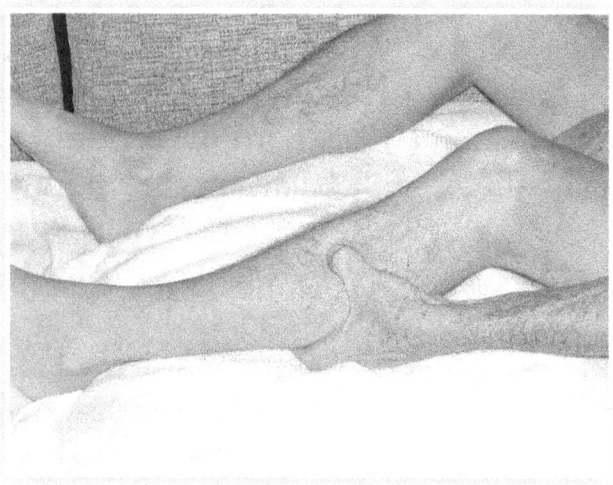

(Figure 1)

When I finally accepted the fact that the numbness in my foot was caused by injuries to my foot, ankle, and lower leg, I began palpating those areas. I found three areas in my lower leg that were extremely painful (Figures 1, 2, and 3), and then iced these areas (Figures 4, 5, 6 and 7). After I'd iced these areas for awhile, I woke up one morning and couldn't put my weight on my left foot, because my Achilles tendon was so painful. I reached down to see where the bulk of the pain was (Figure 8), and began icing it (Figure 9).

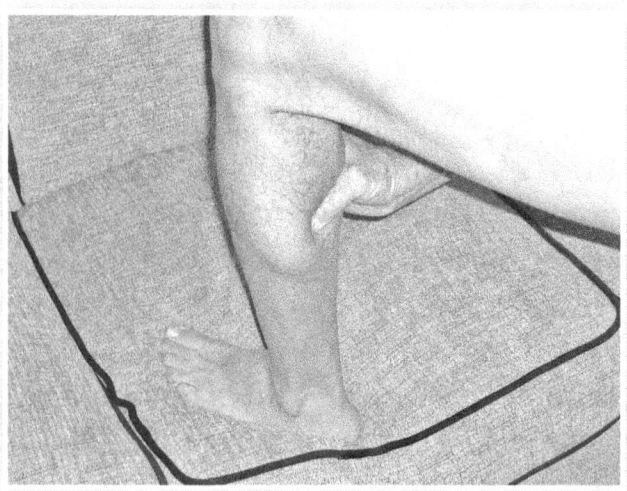

(Figure 2)

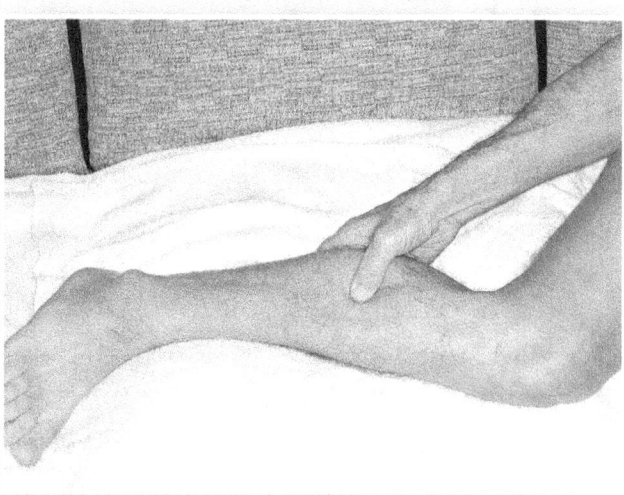

(Figure 3)

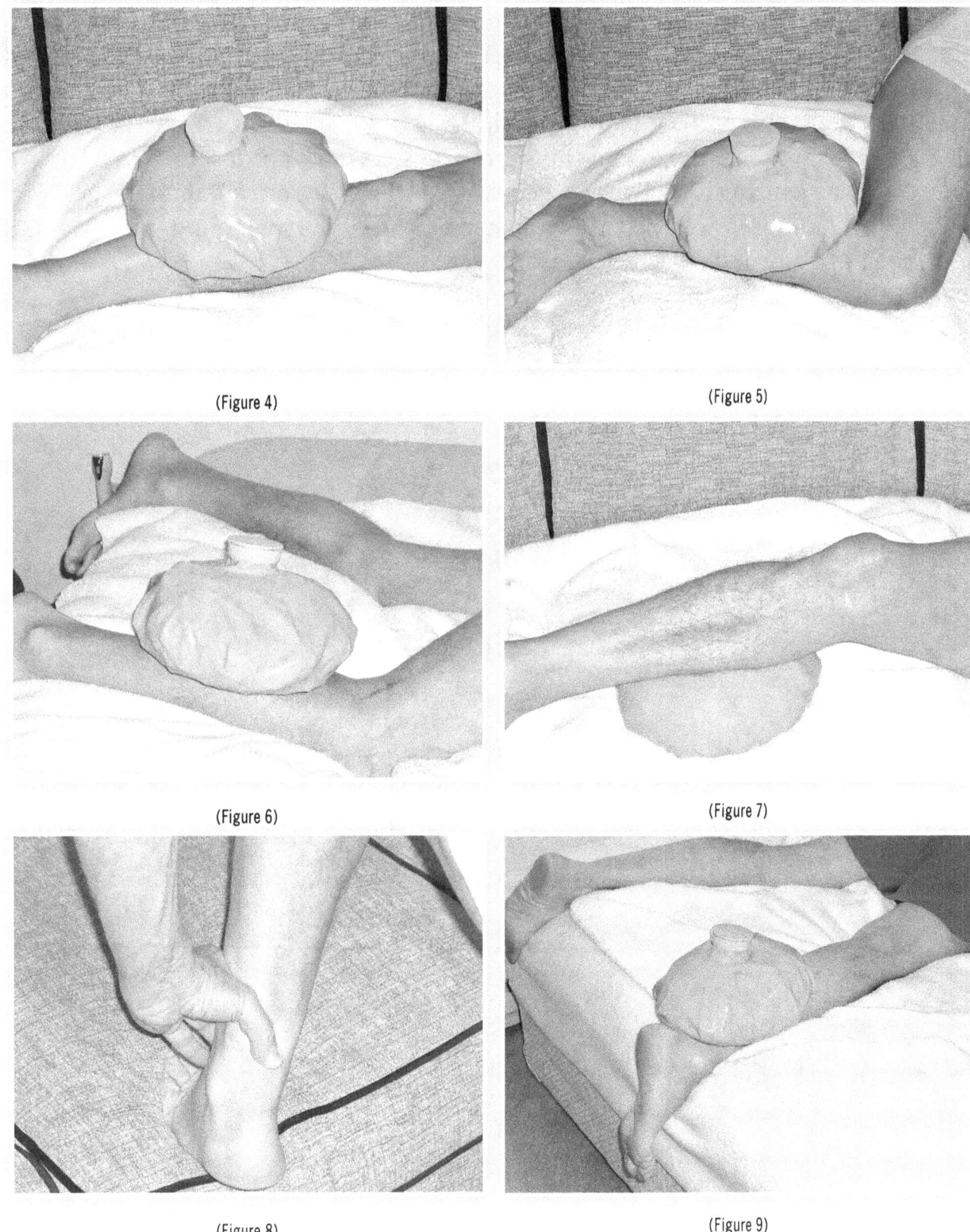

(Figure 4)

(Figure 5)

(Figure 6)

(Figure 7)

(Figure 8)

(Figure 9)

Next, I palpated the foot and ankle and found that the metatarsals and the inside and outside of the foot and ankle were painful (Figures 10, 11, 12, 13, 14, and 15), so I iced those areas as shown in (Figures 16, 17, 18, 19, 20, and 21). I have no idea why the numbness in my left foot didn't show up for years after the injury. I do know this, though: during all the years I was trying to find someone to figure out what was causing it, *not one of them* bothered to take my foot and lower leg in their hands and press on the tissue to check for injury. By icing my lower leg and foot, the numbness has decreased dramatically.

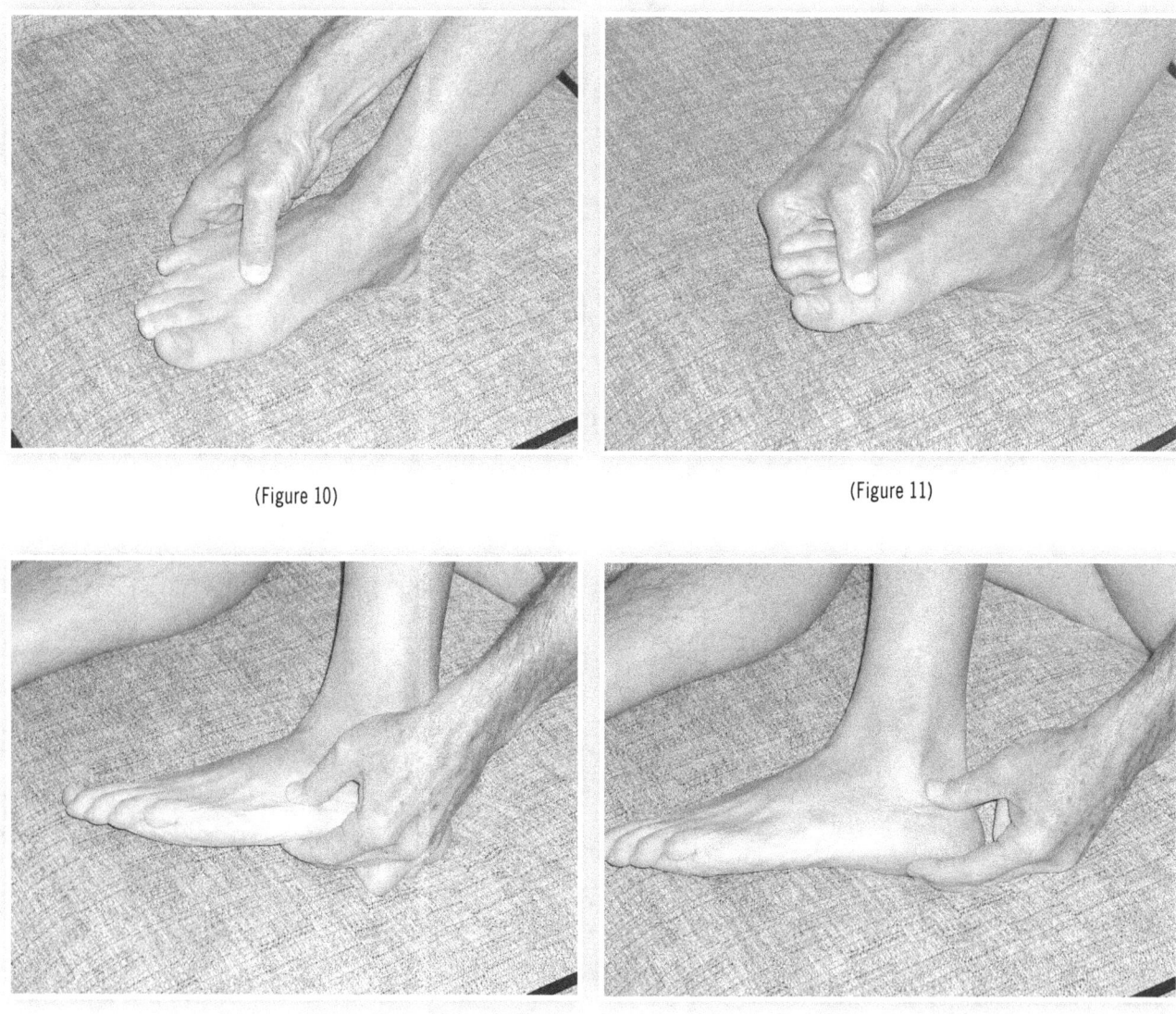

(Figure 10)

(Figure 11)

(Figure 12)

(Figure 13)

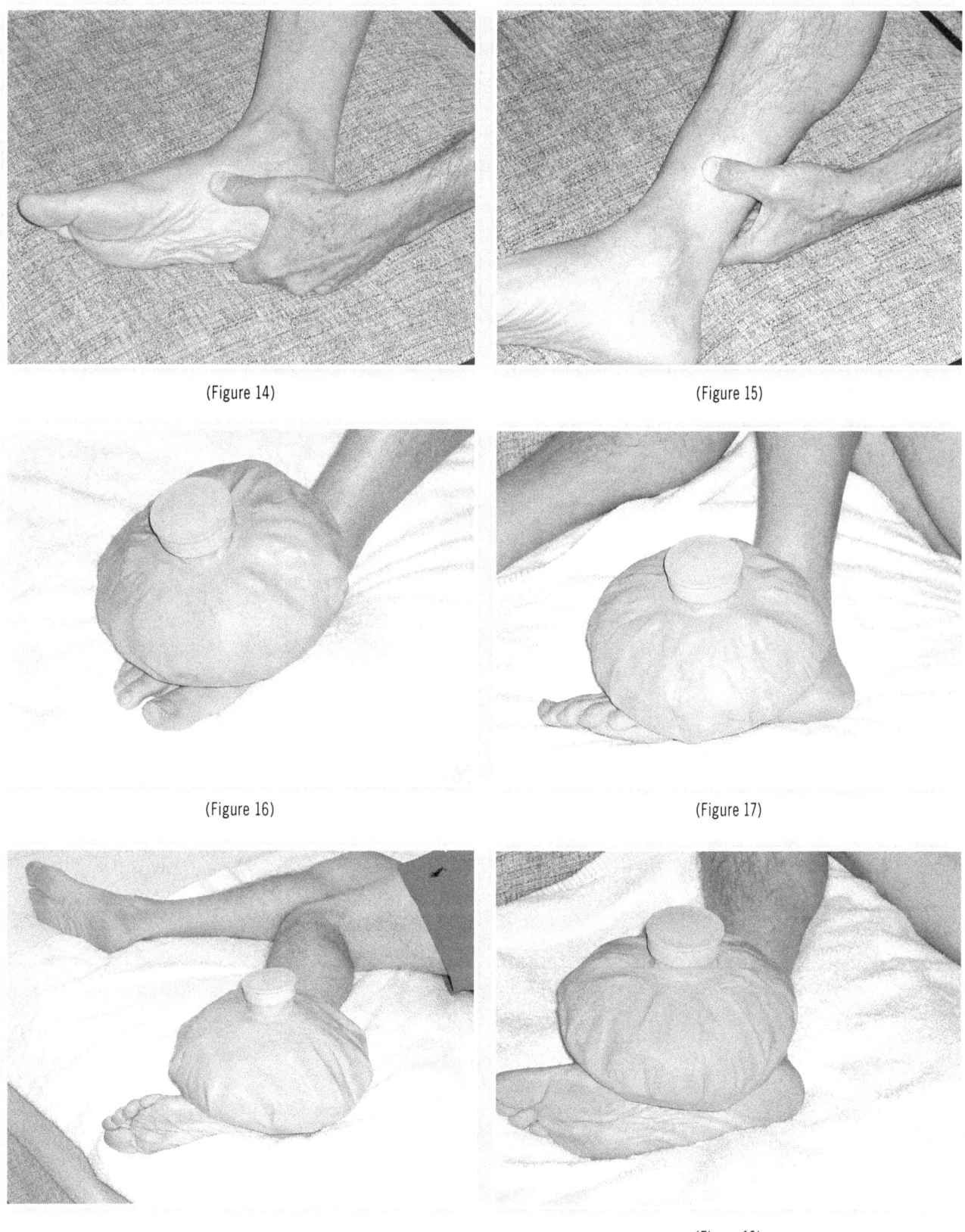

(Figure 14)

(Figure 15)

(Figure 16)

(Figure 17)

(Figure 18)

(Figure 19)

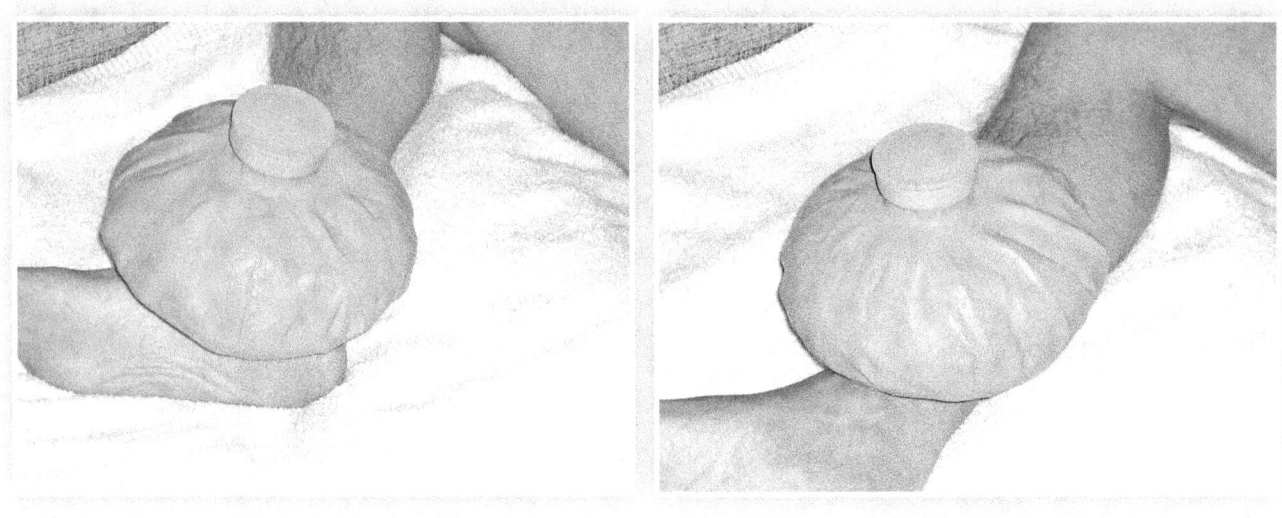

(Figure 20)   (Figure 21)

Section III: The Method

"Laying down on my back w/ the ice bag centered between my eyes, two items struck me the most. One, was the phases and progression of this "healing chill" as it penetrated my forehead. It actually felt as though a physical change was taking place and no one was touching me, no drugs, no electricity or machines. Simply a freeing sensation!

The second was truly feeling movement down in my right thigh, as if something had been released or repositioned....

Change will happen for you as well!"

— Chad Younglove, Santa Monica, CA

Chapter 1. Choosing and Caring for Your Ice Bag

Choosing an ice bag is the most important step you'll take because there's no better way to put yourself off icing than using an ice bag that leaks and gets you and whatever you're icing soaked. If you live in a humid climate, there's going to be a lot of condensation coming off the bag so you'll need to put a couple of folded towels underneath. The best bag I've found is distributed by a chain of pharmacies called Walgreens. It's made of heavy rubber, is manufactured in the USA (see cover photo), and comes with a money-back guarantee. I've used several other brands and they've all leaked sooner or later. I'm not allergic to latex but if you are, you'll need to make sure your bag is non-allergenic.

I use a large or medium-sized ice bag, depending on where the injury is. The medium bag is better for the hands and feet. I keep both in my freezer.

Every bag I've used instructs you to empty it after each use but if you ice often, that's not practical. Getting the ice into the bag is easy. Getting it out is not. If you follow a couple of simple rules, you can safely store your ice bag full of ice and ready to use, in the freezer.

When you're finished icing, drain as much of the water out as possible, refill it, and put it back in the freezer. Wipe the bag dry first so it doesn't stick to the shelf. When you take it out, run it under warm water to melt the sharp edges inside that can cut the rubber. After you've done this, if the ice inside has frozen together, you can safely strike the bag on the kitchen counter to break it up without damaging the bag.

If you take these precautions, your bag should last for months. If your bag springs a leak, you can patch it by filling it with water, finding the leak, and then patching it on the inside. Patch kits are easy to find.

When ice freezes, air is trapped inside, and when the ice melts, the air is released; if the lid on the bag is tightened down, the bag will fill up with air like a balloon. If you leave the lid unscrewed a little, the air can escape, but be careful when the bag is full of water so that you don't dump it on yourself. If you're icing with the bag upside down, screw the lid down tight and when the bag inflates, turn it over, unscrew the lid, and let the air out. Then turn it back over and resume icing.

Chapter 2. A Few Easy Exercises
(and some less easy ones)

Incorporating a few easy exercises into your ice therapy regimen reaps huge rewards. After you've reduced the inflammation and swelling of the connective tissue, and calmed the nerves down, the parts that are out of alignment will realign themselves, and easily fall into place, while you perform these low-impact exercises. With each layer of the "onion of pain" that you peel away, more realignment will take place. Simple tasks like housework and gardening will become easier and you will be more likely to exercise because you'll have less pain holding you back. You will also have more energy at the end of the day.

My encounter with the woman in the Jacuzzi (see Section I, Chapter 3) happened at a time when I had decided to start practicing yoga in the hopes of alleviating some of my pain. Yoga and icing together produced some incredible results. Although I attribute most of the eradication of my pain to ice therapy, there is no doubt that yoga helped to stretch and relax my muscles and had a calming effect on my mind as well. The wonderful part about yoga is that it's a low-impact exercise that doesn't exacerbate your pain by heating up injured tissue. For purposes of relieving pain, please continue to remember that HEAT=BAD, COLD=GOOD. In my opinion, anything that causes *injured* tissue to heat up (other than an occasional hot, relaxing soak in the tub) is not good.

Figures 1 and 2, are the only two exercises that are not optional. It's essential that you perform these maneuvers no matter which body part you are icing, and that you do them each time you get up to drain and reload the ice bag and after you're done. I also perform them right after waking up in the morning and before going to bed at night. These twisting and bending exercises gently coax connective tissue, joints, and their attachments back into place after you've calmed down the nerves and reduced the swelling and inflammation in the tissue. You'll be amazed at the realignments you're going to hear and feel happening in your spine (neck, upper and lower back), ribs, shoulders, and hips when you do these two easy exercises.

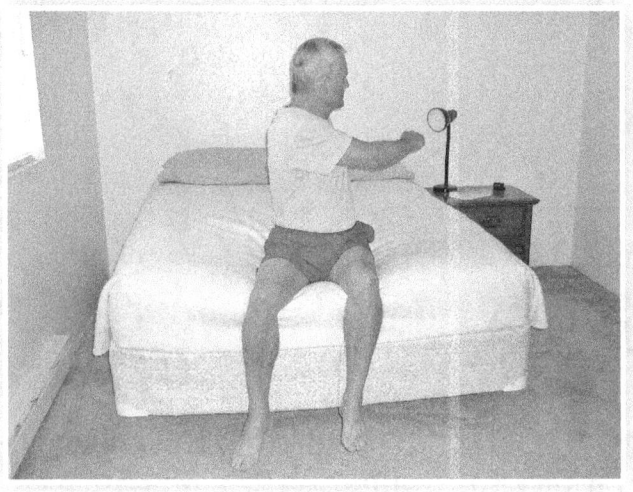

(Figure 1)

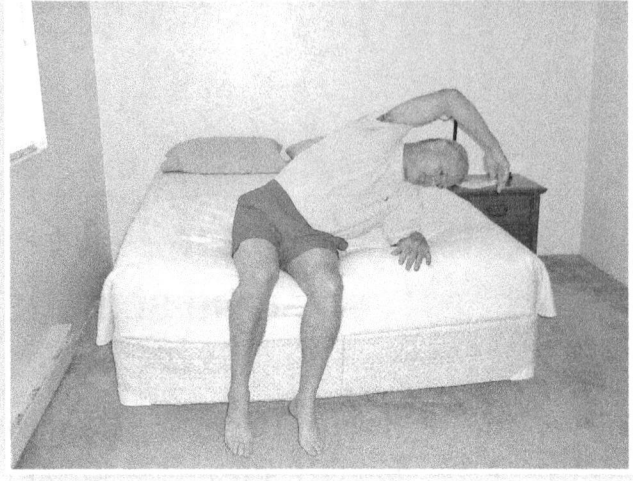

(Figure 2)

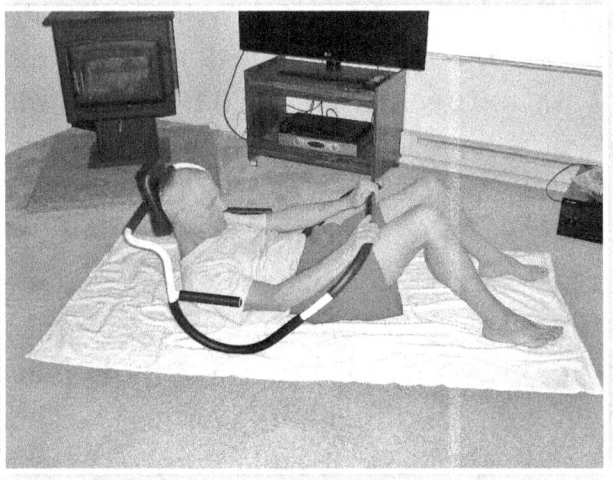

(Figure 3)

Figures 3, 4, and 5, show a sit-up/pull-up machine I call the "gut buster," that I've been using for twelve years. After icing, it helps realign your spine, ribs, and shoulders, and acts as a diagnostic tool by revealing other areas that are still painful. It delivers maximum results using minimum effort and it's great for the abs, too. You can find it in sports stores and online. Some are better than others, so if you're going to do it, you're better off with the more expensive model.

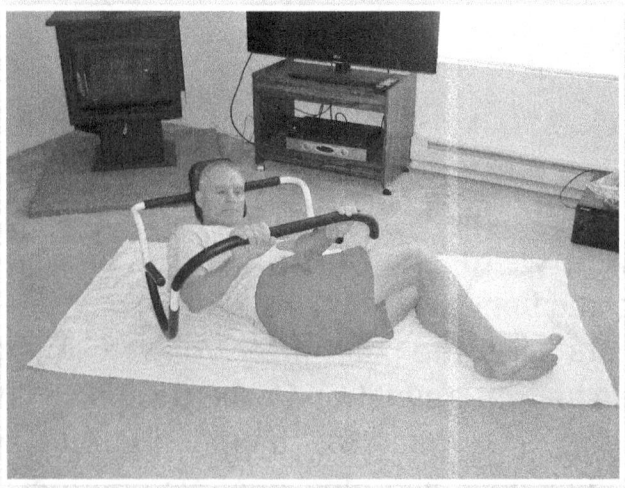

(Figure 4)

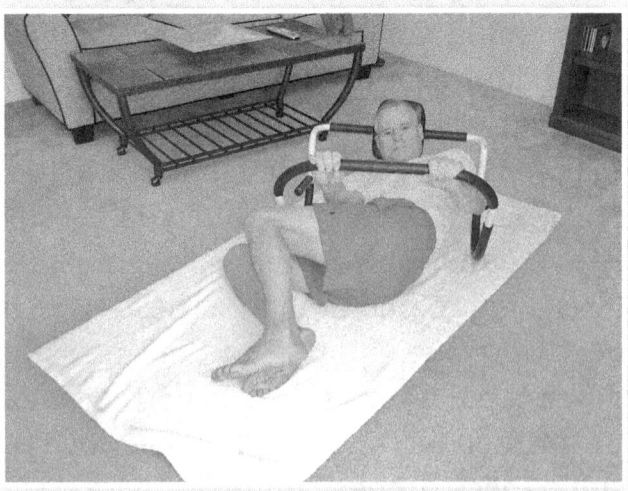

(Figure 5)

Figures 6 and 7, are easy exercises that are great for the spine and hips. I do these before I use the "gut buster."

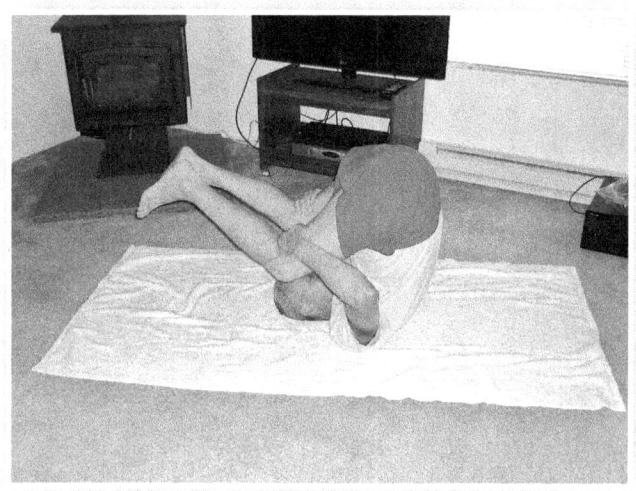

(Figure 6)

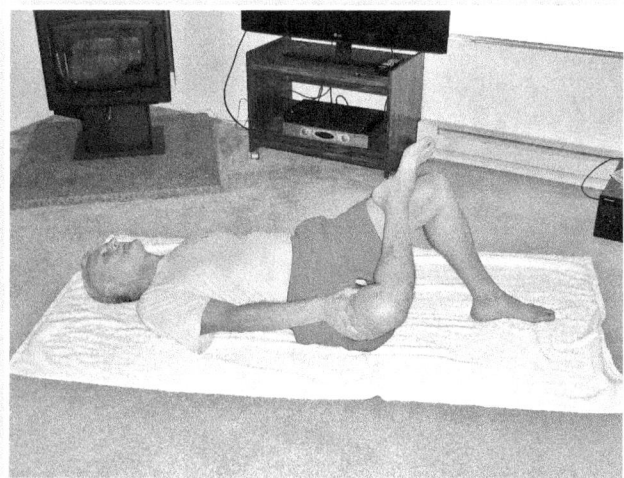

(Figure 7)

Figures 8, 9, 10, and 11, are great for realigning the spine and ribs.

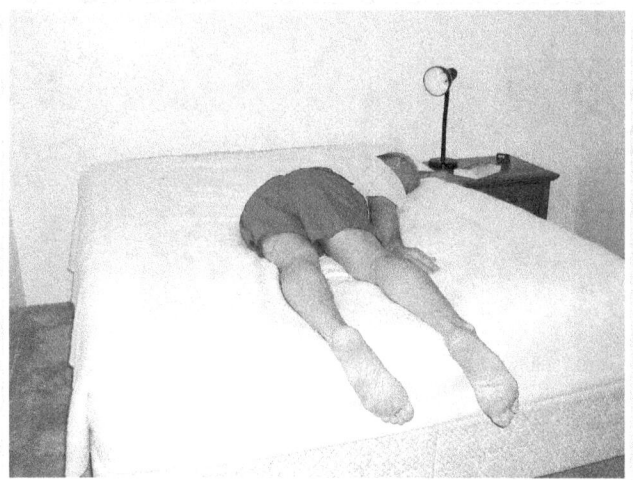

(Figure 8)

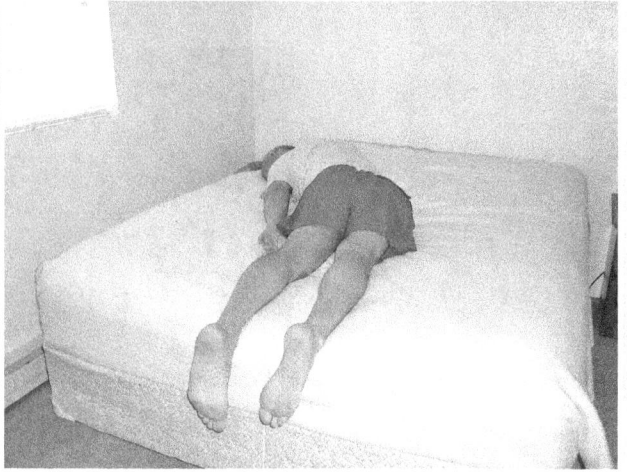

(Figure 9)

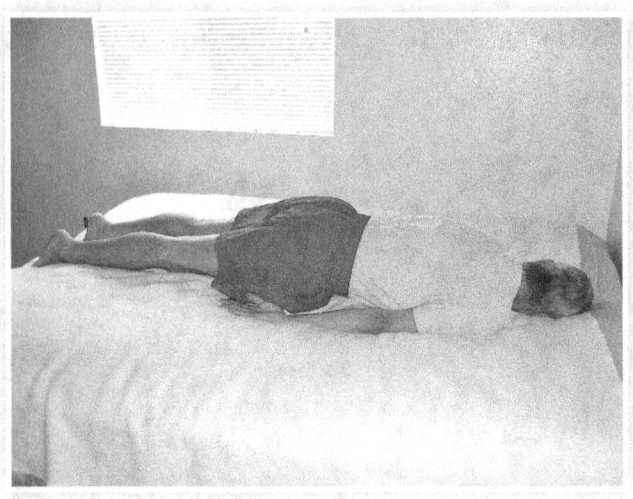

(Figure 10)

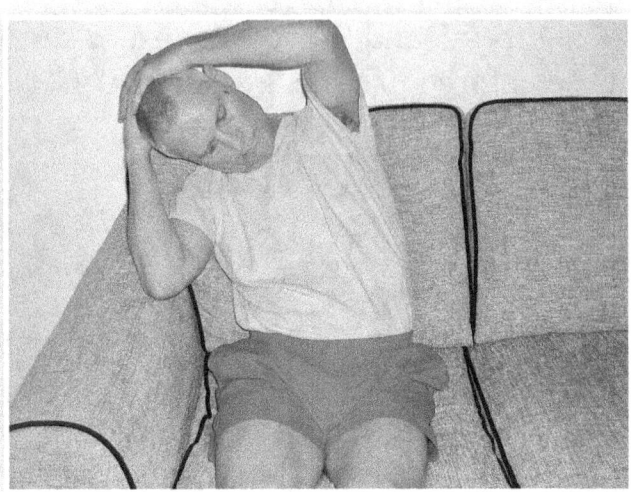

(Figure 11)

Figure 12, is the yoga position called the "cobra." Over the years, when my low back was constantly going out of place, I'd do this exercise to help get it back into place. This pose is good for the entire spine and helps strengthen the lower back and stomach muscles.

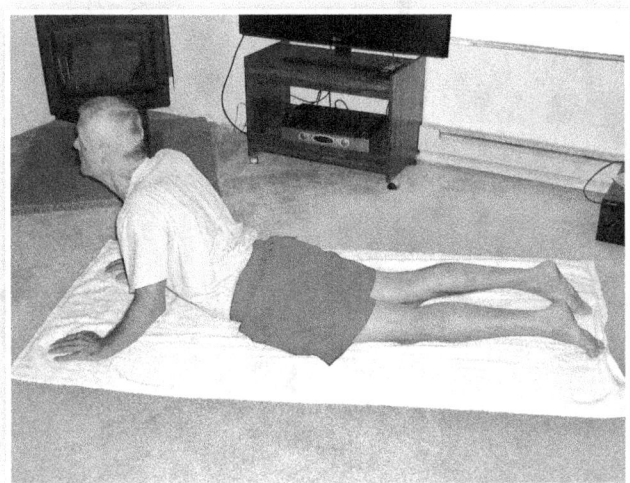

(Figure 12)

Figures 13 (LAO: left anterior oblique) and 14 (RAO: right anterior oblique), are not exercises. They're two positions that you can use to ice the back side of your body and be very comfortable for long periods of time.

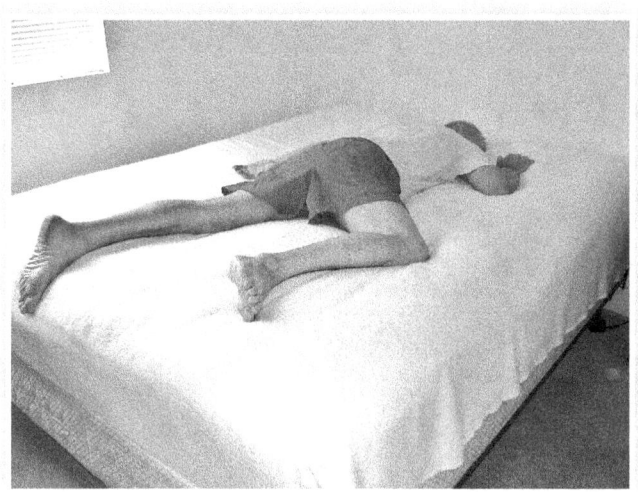

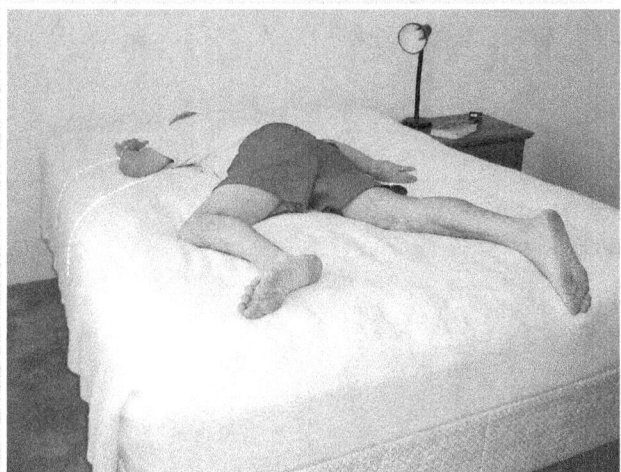

(Figure 13)   (Figure 14)

Be very careful in performing the yoga position called the "plough" for realigning the vertebrae if you have spondylolisthesis (forward displacement of one vertebra over another, usually at the last lumbar vertebra and the first sacral vertebra). (See Chapter 6, Figure 20). If you decide to try this, please have someone stand in as a spotter to support the weight of your legs in case you get into trouble trying to get your legs back up over your head. Don´t expect results until you´ve reduced the swelling and inflammation around the lumbar spine and sacrum/coccyx.

Chapter 3. How to Ice the Pain Away

One of the most important things about icing is that you're comfortable while you're doing it because, if you're not, four hours can be a very long time. Depending on your position, if you're comfortable, you can sleep, read, watch TV, listen to music, or think of a million different things you might plan to do. You'll need privacy while you're icing and it's best to try to keep distractions to a minimum. You'll need pillows for positioning and comfort, and towels to put under you if you live in a humid climate.

Always keep a clock in sight while you're icing. I keep my cell phone handy so that I can see where I'm at in the process and so that I don't have to get up if someone calls me. Keeping track of the time for me is a morale booster and also helps keep me "honest."

I discovered early on that about every forty-five minutes, an event would take place, either at the site of the pain, or in another area of my body, usually on the opposite side to the one I was icing. These events were so regular that I began calling this phenomenon "the forty-five minute rule." It was this "rule" that kept me icing longer and longer and led to the discovery that most old injuries, and the chronic pain associated with them, have to be kept cold for up to four hours before the connective tissue and nerves will respond. A few times I've iced for six hours in order to get an especially stubborn area to let go and begin the healing process. If kept cold long enough, injured tissue will relax, and the nerves will calm down and stop sending pain signals to the brain (or maybe it's the other way around).

During a four-hour period, you will experience five benchmarks: the first at forty-five minutes, the second at one and a half hours, the third at two hours and fifteen minutes, the fourth at three hours, and the fifth at three hours and forty-five minutes. By anticipating these benchmarks, and the events associated with them, you'll find that time passes quickly.

Before you start this process, it's only fair to warn you that where there's injury, there's going to be pain. It hurt when you injured yourself, and it hurts to fix it.

The pain can start in the very first minutes, especially in the hands, feet, elbows, and knees. It may take several tries to work up to hours instead of minutes, but if you keep at it, you'll be successful. Remember, if your injuries are anything like mine, you've been living with them for years, so what's another month or two?

If the area you're icing gets more painful after you take the bag off, run the part under warm (not hot) water and the pain will dissipate quickly. Also, know that it's probable that one icing is not going to fix the problem, especially on old injuries causing chronic pain. You'll probably have to go back several times before you're through, but *please* don't give up. You'll know you're through when you can put the ice bag in place, leave it there for at least three hours, and feel no pain.

It's very important that you don't put anything between the ice bag and your skin. This will diminish the healing power of the cold. Please don't worry that the cold will burn your skin; it's just not possible, as long as you use an ice bag or gel pack designed for this purpose. The only discomfort I've felt is mild soreness for a short time in areas where the skin is very tender, such as the inside of the thigh or the skin on the abdomen.

Remember, ice water is not as cold as ice, so if there's too much water in the bag, it's best to get up, drain the water out, and refill it with ice. In order for ice therapy to be most effective, always try to put the ice bag ON the body part, not UNDER it. But if you absolutely have to, go ahead and put the body part on the bag; just know that the cold may never go deep enough to fix the damage permanently.

There are various reactions that you will experience while you're icing, at the site you're icing or in other areas. They can happen separately or at the same time. These reactions usually occur when the pain in the area you're icing intensifies.

Pain in areas other than the one you're icing usually involve a burning or aching feeling, or both. These reactions in other areas are your body's way of revealing pain you didn't even know you had. I call them "referral pains."

When I first began ice therapy, I experienced so many referral pains that I began writing them down so I didn't forget where they were and I could go back and ice them. That's why I say that the ice bag is a diagnostic tool as well as an instrument of healing.

I wish someone had told me what to expect before I began ice therapy. If you don't know what's coming, some of the reactions can be downright scary. Here are some of them:

- A feeling of electric shock
- A heightened burning sensation
- A mild electric shock and aching in the teeth
- A feeling of immense weight on the body part
- Heart palpitations
- A hollow, breathless feeling in the chest
- The feeling that someone is sticking a needle in your eye
- Sudden sharp pains or aching in the side, groin, or inguinal (the area between mid-body and groin) areas
- Shooting pain in the inner ear
- Vertigo (dizziness)
- Rash
- Nausea
- Itching
- A hollow aching in the sinuses
- A feeling of untwisting in the intestines
- A feeling of untwisting in the muscles
- Intense pain in the joints
- A phenomenon where your body suddenly puts you to sleep when it's ready to "back out" of a painful injury and needs the extra relaxation and non-intervention (from a conscious you) in order to let go

I have been treating myself with ice therapy for years, and the ice bag has been left on the same area for hours. Other than temporary pain, I have experienced only positive and permanent results. If you have any qualms at all about the safety of ice therapy for you, please consult your physician before you begin.

If icing becomes too painful, let the area warm up and try again. If you've convinced yourself that you just don't have the time or can't deal with the pain, consider the alternatives. There are drugs on the market that are specifically designed for people who have connective tissue pain. Carefully listen to, and read, the disclaimers at the end of the commercials that advertise these drugs. We are told that taking the drug might cause suicidal thoughts, internal bleeding, vomiting, stroke or heart attack, or not to take it at all if you're pregnant or have a myriad of other conditions. In addition, we're told not to take some of them with other common drugs; like aspirin. These "disclaimers" blow my mind. Why would anyone risk their life, by taking this stuff?

Section IV: It's Your Adventure Now

"I was involved in a severe auto accident decades ago that threw me out of the car and caused me to slide on the pavement on my head. Harry suggested that I ice the top of my head because I no doubt had hidden pain.

"I iced for three hours and the next day, I had a distinct feeling of euphoria and the feeling that twenty pounds had been lifted from my shoulders. I was ecstatic and decided to continue icing my head to see what other results I got. On subsequent icings, I experienced the phenomenon of being 'put to sleep' to get through the pain. I am still icing this area and getting results."

— Amber Westbrook, Lynnwood, WA

I understand that the two great hurdles standing in the way of ice therapy are pain and time. If there's any activity that gives meaning to the phrase "no pain, no gain", ice therapy is it. And if time is money, consider how much time/money you've spent on treatments that haven't worked. Especially in tough economic times, simple and inexpensive therapy is a terrific benefit.

Unexpectedly, Time magazine supported the case I'm making for ice therapy in its March 7, 2011 issue, "Understanding Pain." In that issue Alice Park writes this about doctors and pain, in her article "Healing the Hurt:"

But recognizing a disease is only a prelude to treating it, and doctors admit that while they're pretty good at relieving the acute pain that occurs immediately after surgery or an injury, they are usually stymied by the chronic kind. The most common complaint doctors hear from their patients is about pain that will not quit, and more than 80% of those people never receive treatment — or at least not an effective one.

Ms. Park goes on to describe how scientists are working on how to alter our DNA to change the way our bodies deal with pain. They admit that this could be dangerous, causing addiction and/or other unknown malfunctions.

Are you convinced yet that spending twelve dollars for an ice bag, and spending a few hours of your time, aren't bad alternatives? Nature has provided us with a gift that we have not fully understood or utilized. It's yours for the taking.

Keep a journal of your own conquest of pain, on this and the following pages.

My Ice Therapy Journal
[In this section put some pages for a journal of the person's conquest]

Afterword

My intention in this guide is not to denigrate anyone but to point out how difficult it is to diagnose and treat pain. In 1998, a U.S. advisory board passed the "patient's bill of rights" that, in part, dealt with issues of pain. I worked in the health care community for nearly twenty years and saw firsthand the caring dedication of my health care co-workers. Medicine and the people that administer it have saved my life on more than one occasion, and for that I'm grateful. The fact still remains, however, that much of what's done to treat and manage pain, can be dangerous to your health and simply doesn't work. Too often, patients are left in pain, worse off than they were before, with the feeling that they are powerless to do anything about it.

Researchers don't put much faith in the results of experiments conducted on just a few lab rats, and I'm painfully (no pun intended) aware that what I've said here will be met with a great deal of skepticism. The only way to prove this new idea for using ice therapy really works is for you to try it yourself.

My goal in writing this book is to do everything I can to help you get well. Please email me at hthmps@yahoo.com if you have any unanswered questions or just want to pick my brain to help find a solution to your problem.

List of Illustrations

Exercises

Head/Scalp

9.   Icing the temple
10.   Palpating sides of upper scalp
11.   Icing side of upper scalp
12.   Palpating brow above eyes
13.   Icing brow above eye
14.   Palpating area behind left ear
15.   Icing area behind left ear
16.   Palpating depression at base of ear
17.   Icing area at base of ear

Cervical Spine (vertebrae of the neck)

1.   Palpating back of neck
2.   Rolled up towels for positioning of neck
3.   Icing back of neck
4.   Palpating side of neck
5.   Icing side of neck
6.   Palpating muscle arising from shoulder
7.   Icing muscle arising from shoulder
8.   Palpating front of neck
9.   Icing front of neck

Shoulder/Upper Arm/Clavicle (collarbone)

1.   Palpating side of shoulder
2.   Palpating side of upper arm
3.   Icing side of shoulder and upper arm
4.   Palpating biceps (the upper arm muscle)
5.   Icing biceps
6.   Palpating front of shoulder
7.   Icing front of shoulder
8.   Palpating back of shoulder
9.   Icing back of shoulder
10.   Palpating clavicle (collarbone)
11.   Icing clavicle
12.   Palpating top of shoulder

5. Palpating attachment of ribs and diaphragm (the muscle and membrane separating the lungs, heart etc. from the lower organs, including intestines)
6. Icing attachment of ribs and diaphragm
7. Palpating bottom of sternum (breast bone)
8. Icing bottom of sternum
9. Palpating top of sternum
10. Icing top of sternum
11. Palpating ribs in armpit
12. Icing ribs in armpit
13. Palpating side of upper ribs
14. Icing side of upper ribs
15. Palpating side of lower ribs
16. Icing side of lower ribs
17. Palpating upper back at level of scapula (shoulder blade)
18. Icing upper back at level of scapula
19. Palpating rib attachments in lower back
20. Icing rib attachments in lower back
21. Palpating upper thoracic spine
22. Icing upper thoracic spine
23. Palpating lower thoracic spine
24. Icing lower thoracic spine
25. "Before" picture of pectus excavatum (depression of the breastbone)
26. "After" picture of pectus excavatum
27. Palpating lateral aspect (outside edge) of scapula
28. Icing lateral aspect of scapula

Lumbar (lower) Spine

1. Palpating the iliac crest (top of the hip bone)
2. Palpating side of lower rib attachments
3. Icing iliac crest and lower ribs
4. Palpating below iliac crest
5. Icing area below iliac crest
6. Palpating lower back muscle
7. Palpating lower rib margin

Hips/Thighs/Knees

Lower Legs/Ankles/Feet

1. Palpating front of lower leg
2. Palpating calves
3. Palpating side of lower leg
4. Icing front of lower leg
5. Icing side of lower leg
6. Icing calf with bag on leg
7. Icing calf with leg on bag
8. Palpating the Achilles tendon (the tendon that attaches to the back of the heel and runs up the back of the lower leg)
9. Icing the Achilles tendon
10. Palpating metatarsals (foot bones)
11. Palpating big toe
12. Palpating side of foot
13. Palpating around anklebone and heel
14. Palpating near instep
15. Palpating inside of leg above ankle
16. Icing top of foot centering on fourth metatarsal
17. Icing outside of foot
18. Icing outside of heel area
19. Icing instep (the bottom part of the arch of the foot)
20. Icing inside of heel area
21. Icing lower leg above ankle

Index

www.ingramcontent.com/pod-product-compliance
Lightning Source LLC
Chambersburg PA
CBHW081358280526
45788CB00009B/2918

*9 7 8 1 4 5 6 3 6 7 2 4 4 *